SpringerBriefs in Computer Science

SpringerBriefs present concise summaries of cutting-edge research and practical applications across a wide spectrum of fields. Featuring compact volumes of 50 to 125 pages, the series covers a range of content from professional to academic.

Typical topics might include:

- A timely report of state-of-the art analytical techniques
- A bridge between new research results, as published in journal articles, and a contextual literature review
- A snapshot of a hot or emerging topic
- An in-depth case study or clinical example
- A presentation of core concepts that students must understand in order to make independent contributions

Briefs allow authors to present their ideas and readers to absorb them with minimal time investment. Briefs will be published as part of Springer's eBook collection, with millions of users worldwide. In addition, Briefs will be available for individual print and electronic purchase. Briefs are characterized by fast, global electronic dissemination, standard publishing contracts, easy-to-use manuscript preparation and formatting guidelines, and expedited production schedules. We aim for publication 8-12 weeks after acceptance. Both solicited and unsolicited manuscripts are considered for publication in this series.

More information about this series at http://www.springer.com/series/10028

Rui Peng • Yan-Fu Li • Yu Liu

Software Fault Detection and Correction: Modeling and Applications

 Springer

Rui Peng
Donlinks School of Economics
& Management
University of Science and Technology
Beijing
Beijing, China

Yan-Fu Li
Department of Industrial & Systems Engineering
Tsinghua University
Beijing, China

Yu Liu
Department of Systems Engineering
and Engineering Management
City University of Hong Kong
Hong Kong, China

ISSN 2191-5768   ISSN 2191-5776   (electronic)
SpringerBriefs in Computer Science
ISBN 978-981-13-1161-1   ISBN 978-981-13-1162-8   (eBook)
https://doi.org/10.1007/978-981-13-1162-8

Library of Congress Control Number: 2018957277

This Springer imprint is published by the registered company Springer Nature Singapore Pte Ltd.
The registered company address is: 152 Beach Road, #21-01/04 Gateway East, Singapore 189721, Singapore

Preface

Software plays an important role in both everyday life and industrial productions. In order to guarantee the reliability of software, a long testing process is usually needed before the software can be finally released to market. During the last few decades, numerous software reliability growth models (SRGMs) have been proposed to characterize the growth of software reliability during testing. These models are very important for management to make related decisions, such as determining the optimal software release time considering both software reliability and total cost.

The models typically can be classified as Markov models, nonhomogeneous Poisson process models, data-driven models, and simulation-based models. These models have been successfully applied into many software projects, especially the NHPP models. However, most traditional models assume that faults can be immediately removed upon detection and thus only construct the fault detection process model. In practice, a detected fault needs to be reported and verified before it can be finally corrected. Some researchers have tried to incorporate the debugging delay into the software reliability modeling process and tried to construct several paired models for both fault detection process and fault correction process. However, these models fail to consider some more practical factors, such as the fault introduction process, the testing resource allocation during testing, and the fault dependency.

In this book, five different works on the modeling of software fault detection and correction processes are introduced, with incorporation of different factors. For each work, the parameter estimation method is introduced, and the applications of the models proposed are illustrated by real dataset. In particular, the optimal release time problem is studied for each work. Chapters 1 and 2 are written by Dr. Yan-Fu LI and Dr. Rui PENG together. Chapters 3, 4, and 7 are written by Dr. Rui PENG. Chapters 5 and 6 are written by Dr. Yu LIU and Dr. Rui PENG together. The following people have also participated in the works included in this book: Prof. Xiangbin Yan and his student Hui Qiu from the University of Science and Technology Beijing, Dr. Qingqing Zhai from Shanghai University, and Chair Professor Min Xie from the City University of Hong Kong. This book is partially supported by the National Natural Science Foundation of China (NSFC) under grant numbers 71531013 and 71671016. The valuable comments from anonymous reviewers and the assistant editor Jane Li are deeply indebted.

Through reading this book, readers should be able to command how the software reliability is modeled, in particular, when both fault detection process and fault correction process are of concern. To better understand the contents of this book, readers are supposed to command the fundamentals of probability theory and stochastic process. This book is especially used for researchers, practitioners, and graduate students in the field of software engineering.

Beijing, China   Rui Peng
Beijing, China   Yan-Fu Li
Hong Kong, China   Yu Liu

Contents

List of Abbreviations

SRGM	Software reliability growth model
"GO model" or "G-O model"	Goel-Okumoto model
MVF	Mean value function
FDP	Fault detection process
FCP	Fault correction process
NHPP	Nonhomogeneous Poisson process
TEF	Testing-effort function
LSE	Least squared error
OLS	Ordinary least square
WLS	Weighted least square
MSE	Mean squared error
RMSE	Root mean squared error
RMSPE	Root mean square prediction error
PDF	Probability density function
CDF	Cumulative distribution function
RADC	Rome Air Development Center
GOS	General order statistics
ETD	Expected time delay
MRE	Mean relative error
FRM	Fault removing matrix
EFRM	Expected FRM
DUCF	Detected but uncorrected faults

List of Notations

$m_d(t)$ or $m(t)$ Mean value function of the fault detection process

$\lambda_d(t)$ or $\lambda(t)$ Fault intensity function for the fault detection process, which equals to $\frac{dm_d(t)}{dt}$ or $\frac{dm(t)}{dt}$

$m_r(t)$ or $m_c(t)$ Mean value function of the fault correction process

$\lambda_r(t)$ or $\lambda_c(t)$ Fault intensity function for the fault correction process

$R(\Delta T | T)$ Software reliability, defined as the probability that no fault occurs during the following ΔT given that the software is released at T

$C(T)$ Total cost function if software is released at time T

Chapter 1
Introduction

This chapter introduces the background of software reliability and the motivation of studying the software fault detection process and fault correction process. Most traditional software reliability models assumed immediate fault correction and studied only the fault detection process. Though some researchers have tried to propose paired models of fault detection process and fault correction process, these models still fail to consider some important practical factors. This book proposes five works on software fault detection and correction processes by considering different factors and illustrating the applications of them with real datasets.

1.1 Reliability

Reliability is usually defined as the probability that a system will operate without failure for a specified time period under specified operating conditions (Xie 1991). With the advancement of technology, systems in safety-critical fields usually contain both hardware components and software components. These types of systems can be found in nuclear plants, aircrafts, and financial institutes. For hardware systems, it may fail if the system elements fail due to aging, rotting, collision, etc. The reliability of hardware element is usually characterized by the failure rate, mean time to failure, etc. These types of information can usually be obtained through making experiments, such as accelerated testing. The reliability of hardware system can be evaluated by considering the reliability of each individual element and the structure of the system. Typical structures of hardware include series structure, parallel structure, series-parallel structure, consecutive k-out-of-n structure, etc. Tools to study the reliability of hardware include fault tree, decision diagrams, universal generating function, simulations, life testing, etc. The reliability of a hardware system is usually enhanced by increasing the reliability of individual system elements and providing redundancy. For systems subject to both internal failures and external impacts, measures can also be taken to protect the system components, such as installing anti-seismic devices, radiation-proof cases, etc.

Different from hardware, software never fails due to aging, rotting, etc. The structure of software codes is also not as clear of the structure of the hardware system. The reasons that can result in software to work improperly are complicated. Say, the faults in software can be due to wrong requirement description of the customers or the misunderstanding of the software developers on the software requirements. Besides, the software faults may be rooted from a wrong design of the software structure. Certainly, the wrong coding during software development is also a main cause of software faults. The longer the software code and the more complicated a software project is, the more faults tend to occur in the software. What makes it worse is that providing multiple identical software versions does not help if the software has some defects leading to software system failure. One way to circumvent this problem is to develop several versions of software independently. However, even so, different versions of software may share some common faults.

1.2   Software Reliability

Unlike the hardware reliability, which already has a relatively mature evaluation methodology, the evaluation of software reliability is still far from mature and integrative. Due to the above reasons, the reliability of software systems is usually much harder to guarantee comparing with that of hardware systems. Even for the systems developed by the National Aeronautics and Space Administration (NASA), the software reliability is usually much lower than the hardware reliability. There-fore, in order to prevent the software from failure, software usually needs to undergo a long testing time before it can be released to the market. During software testing, the software defects may be exposed and get removed. With the removal of software defects, the quality of the software improves, and the reliability of software is increased. In order to characterize the growth of software reliability during the software testing process, many models have been proposed by researchers.

During the last few decades, numerous software reliability models have been proposed to characterize the growth of software reliability during testing. The most famous models include Jelinski-Moranda model, GO model, Yamada S-shaped delayed model, etc. (Lyu 1996; Xie 1991). Whereas the Jelinski-Moranda model and some of its generalizations focus on the inter-arrival time of software faults (Musa et al. 1987), a more recent group of models are the nonhomogeneous Poisson process (NHPP)-based models, which focus on the expected number of detected faults during testing (Kapur et al. 1994; Jha et al. 2009; Wang et al. 2016). Some other software reliability models are constructed using the data-driven approach or the simulation-based approach, including the rate-based simulation approach and the discrete event generation approach. However, most of these models share the assumption that all the faults detected can be immediately removed upon detection, which is unrealistic. In practice, a detected fault usually needs to be reported, verified, before it can be finally removed, and this type of debugging delay may not be negligible. In order to be closer to reality, some researchers have proposed some paired models of fault detection and correction processes. In particular, the fault correction process is modeled as a delayed process of the fault detection process (Xie et al. 2007; Wu et al. 2008). In order to incorporate more practical factors into the paired models, this book introduces some more paired models of fault detection and correction processes, by considering different factors, such as the fault introduction process, the testing resource allocation during testing, the fault dependency, etc. In order for the readers to be able to apply the models proposed in each chapter, the parameter estimation method is introduced for each type of proposed models. The applications of the proposed models are illustrated with real datasets, and the optimal release time policy is studied considering the software reliability and the total software cost.

1.3 Structure of the Book

In Chap. 2, the traditional software reliability models are reviewed. In particular, the models are classified as Markov models, nonhomogeneous Poisson process (NHPP) models, data-driven models, and simulation-based models.

In Chap. 3, some paired models of software fault detection and correction processes are proposed with incorporation of testing-effort function and fault introduction process. Procedures are proposed to detail the selection of the most suitable testing-effort functions and the estimation of the model parameters. The application of the proposed model is illustrated by real dataset, and the optimal release time policy is investigated.

In Chap. 4, the dependency of the faults is incorporated in the software reliability modeling, and some paired models are proposed. Moreover, the models are compared with the models which do not consider the fault dependency. Results show that our models perform much better.

In Chap. 5, a model for software fault detection and correction processes is proposed based on general order statistics. Real dataset on open-source software testing process is used to illustrate the applications.

In Chap. 6, the paired detection and correction models are presented for software with multiple releases. Real data collected from open-sourced software development is used for illustrating the applications of the proposed models.

In Chap. 7, the fault detection and correction processes are modeled considering four types of faults with different detection and correction rates. The advantage of the proposed model is compared with the models considering two types of faults of the same detection rate but different correction rates, two types of faults of the same correction rate but different detection rates, and only one type of fault with the same detection and correction rates.

References

Jha, P. C., Gupta, D., Yang, B., & Kapur, P. (2009). Optimal testing resource allocation during module testing considering cost, testing effort and reliability. *Computers & Industrial Engineering, 57*(3), 1122–1130.

Kapur, P. K., Agarwala, S., Younes, S., & Sinha, A. (1994). On a general imperfect debugging software reliability growth model. *Microelectronics Reliability, 34*(8), 1397–1403.

Lyu, M. R. (1996). *Handbook of software reliability engineering*. New York: McGraw Hill.

Musa, J. D., Iannino, A., & Okumono, K. (1987). *Software reliability, measurement, prediction and application*. New York: McGraw Hill.

Wang, J., Wu, Z., Shu, Y., & Zhang, Z. (2016). An optimized method for software reliability model based on nonhomogeneous poisson process. *Applied Mathematical Modelling, 40*(13–14), 6324–6339.

Wu, Y. P., Hu, Q. P., Xie, M., & Ng, S. H. (2008). Modeling and analysis of software fault detection and correction process by considering time dependency. *IEEE Transactions on Reliability, 56*(4), 629–642.

Xie, M. (1991). *Software reliability modelling*. Singapore: World Scientific.

Xie, M., Hu, Q. P., Wu, Y. P., & Ng, S. H. (2007). A study of the modeling and analysis of software fault-detection and fault-correction processes. *Quality and Reliability Engineering International, 23*, 459–470.

Chapter 2
Classification of Models

This chapter reviews the classical software reliability models proposed in the last few decades. In particular, the models are classified as Markov models, nonhomogeneous Poisson process (NHPP) models, data-driven models, and simulation models. For each type of software reliability models, some typical models are introduced in more detail to illustrate how the model is constructed.

2.1 Software Reliability Models

The software reliability growth modeling has been studied for nearly half a century (Lyu 1996; Xie 1991; Musa et al. 1987; Kapur et al. 1994; Jha et al. 2009; Wang et al. 2016). Among these models, the nonhomogeneous Poisson process (NHPP) models are the most popular (Goel and Okumoto 1979; Yamada et al. 1983). Besides NHPP models, another group of models are the models developed adopting artificial intelligence techniques, such as neural networks, machine learning, etc. (Tian and Noore 2005; Pai and Hong 2006; Hu et al. 2007; Zheng 2009; Lakshmanan and Ramasamy 2015; Wang and Zhang 2018). Some other software reliability models are developed using simulation techniques, such as the rate-based simulation technique and the discrete event generation technique (Hou et al. 2011; Wang et al. 2013; Peng and Shahrzad 2014). Finally, some models are developed based on Markov approach.

2.2 NHPP Models

The nonhomogeneous Poisson process (NHPP)-based software reliability growth models assume that the initial number of faults in software is Poisson distributed. During software testing, the number of remaining faults in software will change with the detection, removal, and sometimes introduction of software faults. The mean value function, which characterizes the expected number of detected or removed faults, is usually used to describe the fault detection or removal process during software testing.

The most fundamental model is the Goel-Okumoto model or GO model in short (Goel and Okumoto 1979). It assumes that the software fault detection rate is constant, and the detected faults are removed immediately upon removal. As the intensity of the fault detection process is proportional to the number of undetected faults in the software, the following relationship holds

$$\lambda_d(t) = \frac{dm_d(t)}{dt} = b(a - m_d(t)) \tag{2.1}$$

where $m_d(t)$ is the mean value function denoting the expected number of detected faults during $(0,t)$, $\lambda_d(t)$ is the fault intensity function characterizing how fast the faults are detected at time t, b is the fault detection rate, and a is the expected number of faults during the beginning of the software testing process. With $m_d(t)$ and $\lambda_d(t)$, the software reliability can be easily formulated. For example, if the software reliability is defined as the probability that software will experience no fault during the next ΔT given that it is released at time T, the software reliability can be given as $R(\Delta T/T) = \exp(-\lambda_d(T)\Delta T)$.

Following GO model, researchers have extended it by incorporating different factors, such as the learning effect, fault coverage, testing-effort function, and imperfect debugging (Yamada et al. 1983; Zeephongsekul 1996; Pham and Zhang 2003; Huang 2005; Inoue and Yamada 2006; Kapur et al. 2008; Li and Pham 2017). However, most of these models have the common assumption that the faults are immediately removed upon detection. In practice, a detected fault needs to be reported and verified before it can be finally removed. In recent years, researchers have started to incorporate the debugging delay into the software reliability modeling and proposed several paired models of fault detection process and fault correction process (Lo and Huang 2006; Xie et al. 2007; Peng and Shahrzad 2014). For instance, Schneidewind (2001) proposed an approach for modeling the fault correction process by using a constant delayed fault detection process. He assumed that the rate of fault correction was proportional to the rate of failure detection. However, since the FCP depends heavily on the FDP, in some applications, the model will underestimate the remaining faults in the code. Later, Xie et al. (2007) and Wu et al. (2008) extended the Schneidewind model to a continuous version by substituting a time-dependent delay function for the constant delay. In Xie et al. (2007), three different types of debugging delay are investigated, being constant debugging delay, time-dependent debugging delay, and exponential distributed random debugging delay. In Wu et al. (2008), more different distributions of debugging delay are investigated, such as gamma distribution and normal distribution. In order to take into account the testing resource allocation among the software testing process and the possible fault introduction effects, Peng and Shahrzad (2014) proposed a framework to analyze the fault detection and correction processes with incorporation of testing-effort function and fault introduction process. Some researchers tried to use a generalized queueing model to study the software fault detection and correction processes. Huang and Huang (2008) and Huang and Hung (2010) applied queueing models to describe the fault detection and correction processes with multiple change points. Gaver and Jocobs (2014) proposed a queue model based on failure mode assumptions.

The fault correcting process is usually assumed to be subject to a stochastic time delay on the FDP with the distribution of $G(t)$. In particular, if the debugging delay is a constant Δt, then the number of faults corrected up to time t equals to the number of faults detected up to time $t - \Delta t$ when $t \geq \Delta t$. Therefore, the intensity function of the fault correction process $\lambda_c^*(t)$ and the mean value function of the fault correction process $m_c^*(t)$ are

$$\lambda_c^* = \begin{cases} \lambda_d(t - \Delta t), & \Delta t \leq t \\ 0, & \Delta t > t \end{cases}, \tag{2.2}$$

$$m_c^* = \begin{cases} m_d(t - \Delta t), & \Delta t \leq t \\ 0, & \Delta t > t \end{cases}. \tag{2.3}$$

In case of a continuous and differentiable $G(t)$, the equation of $\lambda_c(t)$ can be obtained by calculating the expectation of the delayed failure rate (Xie et al. 2007; Wu et al. 2008), which is

$$\lambda_c(t) = E[\lambda_c^*] = \int_0^t \lambda_d(t - x) \cdot g(x) dx, \tag{2.4}$$

$$m_c(t) = \int_0^t \lambda_c(\tau) d\tau = \int_0^t \int_0^\tau \lambda_d(\tau - \Delta) \cdot g(\Delta) d\Delta d\tau$$

$$= \int_0^t \int_0^\tau \lambda_d(x) \cdot g(\tau - x) dx d\tau = \int_0^t \lambda_d(x) G(t - x) dx. \tag{2.5}$$

Since the correction time delay of many software faults can be approximated as exponential distribution (Musa et al. 1987), then the mean value function of fault correction process can be derived as

$$\Delta t \sim \mathrm{Exp}(\mu) \Rightarrow m_c(t) = \begin{cases} a[1 - (1 + rt)e^{-rt}] & , \mu = r \\ a\left[1 - \dfrac{\mu}{\mu - r}e^{-rt} + \dfrac{r}{\mu - r}e^{-\mu t}\right], \mu \neq r \end{cases}. \tag{2.6}$$

To adapt to wider situations, the more flexible Weibull distribution can be used to describe the time delay, which is

$$\Delta t \sim f(x) = \begin{cases} \dfrac{k}{\lambda}\left(\dfrac{x}{\lambda}\right)^{k-1} e^{-(x/\lambda)^k} & x \geq 0 \\ 0 & x < 0 \end{cases} \Rightarrow$$

$$m_c(t) = \int_0^t ab \exp(-\gamma x) \cdot \left\{ 1 - \exp\left(\left[\dfrac{(t-x)}{\lambda}\right]^k\right) \right\} dx. \tag{2.7}$$

where λ is the scale parameter and k is the shaper parameter. It is easy to note that the Weibull time delay degrades to the exponential time delay if the shaper parameter equals to 1.

With incorporation of debugging delay, these models are one step closer to reality. However, these models fail to incorporate more practical factors, such as the fault dependency, the multiple releases of software, the different detection and correction rates of the faults, etc.

2.3 Simulation Models

As it may be mathematically intractable to incorporate into the analytical software reliability models many complex factors, such as the number of debuggers, simulation models are proposed by some researchers. The fault detection and correction processes for software testing can be naturally simulated as a queueing system, where the fault detection is the arrival process and the fault correction is the departure process.

The simulation approaches can be classified as the discrete simulation approach and the rate-based simulation approach. In Peng and Shahrzad (2014), the discrete simulation approach is used to model the fault detection and correction processes. In particular, the number of debuggers is assumed to be finite, and a queueing system model is constructed with Simulink toolbox in MATLAB. Moreover, the debuggers are divided into a few groups according to their debugging capabilities, as shown in Fig. 2.1.

Fig. 2.1 The queueing system model for fault detection and correction processes

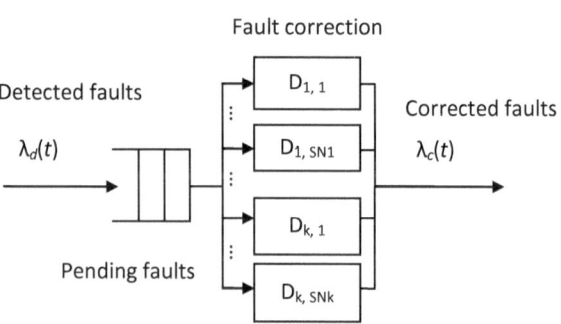

In Fig. 2.1, $\lambda_d(t)$ is the fault intensity function for the fault detection process, which is simulated using ordered statistics generated from the GO model (Goel and Okumoto 1979). Different types of debuggers are simulated with a queue of different types of servers, where k is the total number of types, SN_i is the number of debuggers of type 1, and $D_{i,j}$ is jth debugger of type i. In order to accelerate the fault correction, it is assumed that the detected fault will choose the vacant debugger of the best debugging capability as long as there is any vacant debuggers. Different fault correction time distributions are investigated for the debuggers, where the fault correction time is generated by using the inverse transformation of the cumulative distribution function of the fault correction time. Based on $\lambda_d(t)$ and fault correction time, the intensity function for the fault correction process $\lambda_c(t)$ can be obtained. In addition, the optimal release time together with the optimal staffing level is investigated. Later, Peng and Liu (2017) further extended the model to incorporate the different influence of different types of debuggers on the fault detection rate.

Rate-based simulation, whose theoretical foundation is pure birth nonhomogeneous continuous time Markov chain, is expected to consume less computational resource than discrete event generation for complex process simulation. With simulation, it is possible to incorporate into the software reliability models many factors that would be mathematically intractable for analytical models. Tausworthe and Lyu first attempted to develop a general rate-based simulation technique to relax certain unreasonable assumptions of SRGMs (Lyu 1996; Tausworthe and Lyu 1996). Gokhale and Lyu (2005) proposed an approach to structure-based software reliability analysis based on the framework of rate-based simulation because this framework can be used to analyze the stochastic failure process which may be analytically intractable. Later, Gokhale et al. (2006) incorporated fault debugging activities into software reliability models based on the same framework. In recent years, Lin and Huang (2009) incorporated the concept of queueing model into the rate-based simulation framework based on which they analyzed the best staffing level and associated cost for debugging activities in software development. In order to approximate reality more closely, Lin (2011) further enhanced the rate-based queueing simulation model by considering the effect of imperfect debugging on fault detection and correction processes.

2.4 Data-Driven Models

Restricted by mathematical tractability, it is difficult for analytical software reliability models to incorporate many complex factors. Therefore, some researchers tried to use data-driven approach to study the reliability of software. Data-driven approach regards the software failure process as a time series and tries to recognize the inherent patterns of the process from historical data. Based on the recognized patterns, software reliability prediction can be made. Most of data-driven software reliability models are based on multiple-delayed-input single-output architecture. Tian and Noore (2005) predicted the cumulative software failure time based on an

Fig. 2.2 The recurrent
network for prediction of the
fault detection and
correction numbers

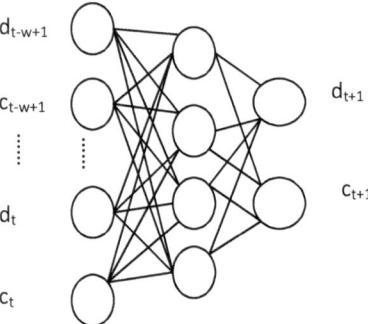

evolutionary neural network technique. Hu et al. (2007) adopted the recurrent neural
network to study and predict the number of detected and corrected software faults. In
particular, a robust recurrent neural network modeling process for software fault
detection and correction prediction is proposed. The approach uses the numbers of
detected faults and corrected faults during a past period of time to predict the
numbers of detected faults and corrected faults at the next time point. This is
achieved by training the neural network as shown in Fig. 2.2, where d_j and c_j
represent the number of detected faults and the number of corrected faults at time
point j.

Su and Huang (2007) incorporated the dynamic weighted combinatorial models
into the neural network-based approach for software reliability estimation. Some
other works used support machine vector approach to predict the software faults (Pai
and Hong 2006; Yang et al. 2010).

2.5   Markov Models

Markov chain has wide applicability in modeling the state transitions of complex
systems, where the transition time has exponential distribution. When the state
transition time is exponential, it endows the system Markov property, or memoryless
property, that is, the state of the system at one time point is only related to the state of
the system at the last time point no matter the states experienced before the last time
point. Jelinski-Moranda model is the most famous model among the early developed
software reliability models (Jelinski and Moranda 1972). In particular, it assumes
that the fault detection rate is constant between the detection of any two faults, and it
is proportional to the number of remaining faults in the software. Without consideration of the debugging delay and the fault introduction effects, it is easy to see that

$$f(x_i|t_{i-1}) = \eta(N - i + 1)\exp(-\eta(N - i + 1)t). \qquad (2.8)$$

where N is the total number of faults in the software, t_{i-1} is the time when the $(i-1)$th fault is detected, x_i is the time for the detection of the ith fault counting from t_{i-1}, η is the rate for each remaining fault detected, and $f(x_i|t_{i-1})$ is the probability density function for x_i given t_{i-1}. It can be seen that $f(x_i|t_{i-1})$ observes exponential distribution, as the fault detection rate of the software faults is assumed to be constant. Different from the NHPP models which usually need to collect the actual numbers of detected and/or corrected faults during different time intervals for model parameter estimation, one needs to collect the inter-arrival time of fault detections in order to apply the Jelinski-Moranda model. A typical way is to calculate the likelihood that the fault inter-arrival times are as observed and then find out the combinations of model parameters which maximize the likelihood function. Once the parameters in (2.8) are estimated, it is easy to obtain the mean value of detected faults at time t as

$$m(t) = N(1 - \exp(-\eta t)). \tag{2.9}$$

Following Jelinski-Moranda model, some more Markov models have been proposed to adapt to different situations. Whittaker et al. (2000) adopted a Markov chain-based approach to predict the reliability of multi-build software. Extensions on the Markov models have also been conducted by Gokhale et al. (2004) in a systematic way. Wei et al. (2011) used Markov approach to analyze the reliability of hierarchical software. Landon et al. (2013) used a Markov-modulated Poisson model to study the software reliability, where a latent Markov process is used to govern the intensity rate of a Poisson process model for software failures.

2.6 Summary

Numerical software reliability growth models have been proposed during the last few decades, which can be mainly classified as Markov models, NHPP models, simulation models, and data-driven models. The Markov model mainly focuses on the detection time of the next fault, whereas the other three types focus more on the numbers of detected and corrected faults. Among all the models, the NHPP models are most popular due to its flexibility and its successful application in practice. The most famous NHPP models for fault detection process are GO model and Yamada delayed S-shaped model. In order to be more realistic, some paired models of fault detection process and fault correction process are proposed in Xie et al. (2007), Lo and Huang (2006), Wu et al. (2008), etc. However, trying to incorporate more complicated factors into the NHPP models is challenging due to mathematical tractability. In order to be able to incorporate factors such as the staffing level, researchers have tried to simulate the software testing process using either discrete event generator or rate-based simulation approach. With the development of neural networks and support vector machine, some data-driven models have also been developed for prediction of the software fault detection and correction process.

References

Gaver, D. P., & Jocobs, P. A. (2014). Reliability growth by failure mode removal. *Reliability Engineering & System Safety, 130*, 27–32.

Goel, A. L., & Okumoto, K. (1979). Time-dependent error-detection rate model for software reliability and other performance measures. *IEEE Transactions on Reliability, R28*(1979), 206–211.

Gokhale, S., & Lyu, M. R. (2005). A simulation approach to structure-based software reliability analysis. *IEEE Transactions on Software Engineering, 31*(8), 643–656.

Gokhale, S. S., Lyu, M. R., & Trivedi, K. S. (2004). Analysis of software fault removal policies using a non-homogeneous continuous time Markov chain. *Software Quality Journal, 12*(3), 211–230.

Gokhale, S., Lyu, M. R., & Trivedi, K. S. (2006). Incorporating fault debugging activities into software reliability models: A simulation approach. *IEEE Transactions on Reliability, 55*(2), 281–292.

Hou, C. Y., Cui, G., & Liu, H. W. (2011). Rate-based component software reliability process simulation. *Journal of Software, 22*(11), 2749–2759.

Hu, Q. P., Xie, M., Ng, S. H., & Levitin, G. (2007). Robust recurrent neural network modeling for software fault detection and correction prediction. *Reliability Engineering & System Safety, 92*(3), 332–340.

Huang, C. Y. (2005). Performance analysis of software reliability growth models with testing-effort and change-point. *Journal of Systems and Software, 76*(2), 181–194.

Huang, C. Y., & Huang, W. C. (2008). Software reliability analysis and measurement using finite and infinite server queueing models. *IEEE Transactions on Reliability, 57*(1), 192–203.

Huang, C. Y., & Hung, T. Y. (2010). Software reliability analysis and assessment using queueing models with multiple change-points. *Computers & Mathematics with Applications, 60*(7), 2015–2030.

Inoue, S., & Yamada, S. (2006). Discrete software reliability assessment with discretized NHPP models. *Computers & Mathematics with Applications, 51*(2), 161–170.

Jelinski, Z., & Moranda, P. B. (1972). Statistical computer performance evaluation. *Software Reliability Research*, 465–484.

Jha, P. C., Gupta, D., Yang, B., & Kapur, P. (2009). Optimal testing resource allocation during module testing considering cost, testing effort and reliability. *Computers & Industrial Engineering, 57*(3), 1122–1130.

Kapur, P. K., Agarwala, S., Younes, S., & Sinha, A. (1994). On a general imperfect debugging software reliability growth model. *Microelectronics Reliability, 34*(8), 1397–1403.

Kapur, P. K., Goswami, D. N., Bardhan, A., & Singh, O. (2008). Flexible software reliability growth model with testing effort dependent learning process. *Applied Mathematical Modelling, 32*(7), 1298–1307.

Lakshmanan, I., & Ramasamy, S. (2015). An artificial neural-network approach to software reliability growth modeling. *Procedia Computer Science, 57*, 695–702.

Landon, J., Ozekici, S., & Soyer, R. (2013). A Markov modulated Poisson model for software reliability. *European Journal of Operational Research, 229*(2), 404–410.

Li, Q., & Pham, H. (2017). NHPP software reliability model considering the uncertainty of operating environments with imperfect debugging and testing coverage. *Applied Mathematical Modelling, 51*, 68–85.

Lin, C. T. (2011). Analyzing the effect of imperfect debugging on software fault detection and correction processes via a simulation framework. *Mathematical and Computer Modelling, 54* (11–12), 3046–3064.

Lin, C. T., & Huang, C. Y. (2009). Staffing level and cost analyses for software debugging activities through rate-based simulation approaches. *IEEE Transactions on Reliability, 58*(4), 711–724.

Lo, J., & Huang, C. (2006). An integration of fault detection and correction processes in software reliability analysis. *Journal of Systems and Software, 79*(9), 1312–1323.

Lyu, M. R. (1996). *Handbook of software reliability engineering*. New York: McGraw Hill.

Musa, J. D., Iannino, A., & Okumono, K. (1987). *Software reliability, measurement, prediction and application*. New York: McGraw Hill.

Pai, P. F., & Hong, W. C. (2006). Software reliability forecasting by support vector machines with simulated annealing algorithms. *Journal of Systems and Software, 79*(6), 747–755.

Peng, R., & Liu, J. (2017). Simulated software testing process considering debuggers with different detection and correction capabilities. *International Journal of Performability Engineering, 13*(3), 319–321.

Peng, R., & Shahrzad, F. R. (2014). *Simulation of software fault detection and correction processes considering different skill levels of debuggers*. In Proceedings of the 2014 I.E. 20th Pacific Rim International Symposium on Dependable Computing, November 18–21, 2014, 157–158.

Pham, H., & Zhang, X. (2003). NHPP software reliability and cost models with testing coverage. *European Journal of Operational Research, 145*(2), 443–454.

Schneidewind, N. F. (2001). *Modelling the fault correction process*. In Proceedings 12th international symposium on software reliability engineering, pp. 185–190.

Su, Y. S., & Huang, C. Y. (2007). Neural-network-based approaches for software reliability estimation using dynamic weighted combinational models. *Journal of Systems and Software, 80*(4), 606–615.

Tausworthe, R. C., & Lyu, M. R. (1996). A generalized technique for simulating software reliability. *IEEE Software, 13*(2), 77–88.

Tian, L., & Noore, A. (2005). Dynamic software reliability prediction: An approach based on support vector machines. *International Journal of Reliability, Quality, and Safety Engineering, 12*(4), 309–321.

Wang, J., & Zhang, C. (2018). Software reliability prediction using a deep learning model based on the RNN encoder–decoder. *Reliability Engineering & System Safety, 170*, 73–82.

Wang, Q., Lu, Y., Wu, L., & Wei, Z. (2013). Software reliability simulation research containing fault recovery. *Journal of System Simulation, 25*(5), 887–893.

Wang, J., Wu, Z., Shu, Y., & Zhang, Z. (2016). An optimized method for software reliability model based on nonhomogeneous Poisson process. *Applied Mathematical Modelling, 40*(13–14), 6324–6339.

Wei, Y., Wang, L., & Wang, M. (2011). Software reliability analysis of hierarchical architecture based on Markov model. *Procedia Engineering, 15*, 2857–2861.

Whittaker, J., Rekab, K., & Thomason, M. G. (2000). A Markov chain model for predicting the reliability of multi-build software. *Information and Software Technology, 42*(12), 889–894.

Wu, Y. P., Hu, Q. P., Xie, M., & Ng, S. H. (2008). Modeling and analysis of software fault detection and correction process by considering time dependency. *IEEE Transactions on Reliability, 56* (4), 629–642.

Xie, M. (1991). *Software reliability modelling*. Singapore: World Scientific.

Xie, M., Hu, Q. P., Wu, Y. P., & Ng, S. H. (2007). A study of the modeling and analysis of software fault detection and fault-correction processes. *Quality and Reliability Engineering International, 23*, 459–470.

Yamada, S., Ohba, M., & Osaki, S. (1983). S-shaped reliability growth modeling for software error detection. *IEEE Transactions on Reliability, 32*, 475–484.

Yang, B., Li, X., Xie, M., & Tan, F. (2010). A generic data-driven software reliability model with model mining technique. *Reliability Engineering & System Safety, 95*(6), 671–678.

Zeephongsekul, P. (1996). Reliability growth of a software model under imperfect debugging and generation of errors. *Microelectronics Reliability, 36*(10), 1475–1482.

Zheng, J. (2009). Predicting software reliability with neural network ensembles. *Expert Systems with Applications, 36*(2), 2116–2122.

Chapter 3
TEF Dependent Software FDP and FCP Models

This chapter incorporates the fault introduction process and the testing resource allocation into the modeling of software fault detection and correction processes. Several paired models for fault detection process and fault correction process are constructed by considering different assumptions of correction effort. The applications of the models are illustrated with real dataset, and the optimal software release time is studied.

3.1 Introduction

As software is becoming more and more widely used, both the functionality and the correctness of software are of great concern. In order to ensure high reliability, testing is usually conducted, during which faults in software manifest by causing failures and can be detected and removed by debuggers (Hu et al. 2007; Chiu et al. 2008; Kang et al. 2009). On the other hand, it is almost impossible to make bug-free software even though scientific and disciplined development practices are followed. During the last 30 years, many software reliability growth models (SRGMs) have been proposed as a tool to track the reliability growth trend of software testing process (Kapur et al. 2008; Chang and Liu 2009; Yang et al. 2010; Okamura et al. 2013). SRGMs are very useful in the sense that they can help management making critical decisions, such as testing resource allocation and the determination of software release time (Kim et al. 2009; Jha et al. 2009; Jain and Gupta 2011).

Testing consumes a large amount of resources, such as manpower and CPU hours, which are usually not constantly allocated during testing phase. The function that describes how testing resources are distributed is usually referred to as testing-effort function (TEF), and it has been incorporated into software reliability studies by some researchers (Stikkel 2006; Kapur et al. 2008, 2009). Yamada et al. (1992)

pointed out that the TEF could be described by a Weibull-type distribution, which actually includes exponential curve, Rayleigh curve, and Weibull curve. Weibull-type curve can well fit most data and is often used in the field of software reliability modeling (Huang 2005). Logistic TEF is used instead of Weibull-type TEF by some researchers and appeared to be fairly accurate in describing the consumption of testing effort (Parr 1980; Demarko 1982; Huang and Kuo 2002).

Generally a detected fault cannot be corrected immediately, and the time required to correct a detected fault is usually called debugging lag/delay. The idea of modeling fault correction process (FCP) was first proposed in Schneidewind (1975), in which it was modeled as a separate process following fault detection process (FDP) with a constant time lag. Based on this framework, Xie et al. (2007) and Wu et al. (2008) proposed several paired FDP and FCP models through incorporating other variants of debugging delay. Later, Hwang and Pham (2009) developed a generalized NHPP model considering quasi-renewal time-delay fault removal. Jia et al. (2010) proposed a Markovian software reliability model considering fault correction process. However, the influence of testing effort on debugging lag is not considered in these papers. Intuitively, the time needed to correct a detected fault, or the debugging lag, tends to be shorter if more testing effort is allocated during the period between detection and correction of the fault. Thus it is more reasonable to incorporate testing-effort function into the modeling framework on both FDP and FCP.

Moreover, debugging process is usually far from perfect, and actually many faults encountered by customers are those introduced during debugging (Zhang et al. 2003; Shyur 2003; Gokhale et al. 2006; Pievatolo et al. 2012). It is essential to incorporate imperfect debugging into FDP and FCP models (Xie and Yang 2003; Cai et al. 2010; Kapur et al. 2011).

In this chapter, a framework is proposed to develop testing-effort dependent FDP and FCP models with the consideration of imperfect debugging. The rest of this chapter is organized as follows. In Sect. 3.2, a framework is proposed to obtain testing-effort dependent paired FDP and FCP models with the consideration of fault introduction. Besides, several specific models are derived based on different assumptions of fault introduction and the correction effort. In addition, several commonly used testing-effort functions are reviewed. In Sect. 3.3, the parameter estimation method is investigated. In Sect. 3.4, an illustrative example is presented. The optimal release policy under different criteria is studied in Sect. 3.5.

3.2 The Modeling Process

The expected total number of faults at time t is denoted by the fault content rate function $a(t)$, which is the sum of the number of initial faults in the software a ($=a(0)$) and the number of faults introduced during time interval $[0, t)$. We use $w(t)$ to denote the time-dependent testing-effort rate and $W(t)$ to denote the cumulative testing effort consumed till time t.

3.2.1   *FDP Model*

Mean value function $m_d(t)$ is used to depict the expected number of faults detected till time t and $\lambda_d(t) = \frac{dm_d(t)}{dt}$ is used to denote the fault intensity function. The number of faults detected during time interval $[t, t+\Delta t)$ by current testing-effort expenditure is usually assumed to be proportional to the number of remaining faults at time t (Lin and Huang 2008). Hence we have

$$\lambda_d(t) = \frac{dm_d(t)}{dt} = b(t) \cdot w(t) \cdot (a(t) - m_d(t)) \tag{3.1}$$

where $b(t)$ is the current fault detection rate per unit of testing effort at time t and $w(t)$ is the current testing-effort expenditure at time t. Substituting the marginal condition $m_d(0){=}0$ into (3.1) gives

$$m_d(t) = a(t) - a \cdot \exp\left\{ -\int_0^t b(x)w(x)dx \right\}$$
$$-\exp\left\{ -\int_0^t b(x)w(x)dx \right\} \cdot \int_0^t a'(x)\exp\left\{ \int_0^x b(y)w(y)dy \right\}dx \tag{3.2}$$

where $a'(x) = \frac{da(x)}{dx}$. Various $m_d(t)$ can be derived based on different assumptions of $a(t)$, $b(t)$, and $w(t)$. $\lambda_d(t)$ can be obtained by substituting (3.2) into the right-hand side of (3.1) as

$$\lambda_d(t) = \frac{dm_d(t)}{dt}$$
$$= a \cdot b(t) \cdot w(t) \cdot \exp\left\{ -\int_0^t b(x)w(x)dx \right\}$$
$$\cdot \left(1 + \int_0^t \frac{a'(x)}{a}\exp\left\{ \int_0^x b(y)w(y)dy \right\}dx \right) \tag{3.3}$$

3.2.2   *FCP Model*

Mean value function $m_r(t)$ is used to denote the expected number of faults removed till time t and $\lambda_r(t) = \frac{dm_r(t)}{dt}$ is used to denote the fault removal intensity function. Since a removed fault must first be detected, FCP can be modeled as a separate process following FDP with a debugging delay. For convenience of discussion, the testing effort consumed during the period from detection of a fault to the final removal of the fault is termed as correction effort of the fault. Generally correcting different faults requires different amounts of testing resources; hence correction effort can be modeled as a random variable with probability density function (PDF) and the cumulative distribution function (CDF) denoted as $f(x)$ and $F(x)$.

Thus it can be obtained that

$$m_r(t) = \int_0^t \lambda_d(y)F(W(t) - W(y))dy \tag{3.4}$$

where $F(W(t) - W(y))$ is the probability that the fault detected at y is corrected before t.

Different $m_r(t)$ can be derived based on $m_d(t)$ and different $f(x)$. Furthermore, we have

$$\begin{aligned}
\lambda_r(t) &= \int_0^t \lambda_d(y)f(W(t) - W(y))w(t)dy \\
&= \int_0^{W^*(t)} \lambda_d(W^{-1}(W(t) - x))f(x)w(t)d(W^{-1}(W(t) - x)) \\
&= \int_0^{W^*(t)} \lambda_d(W^{-1}(W(t) - x))f(x)w(t)\frac{dx}{w(W^{-1}(W(t) - x))}
\end{aligned} \tag{3.5}$$

Different $m_r(t)$ can be derived based on $m_d(t)$ and different $f(x)$.

3.2.3   Some Specific Models

Fault detection rate function $b(t)$ is usually assumed to be constant, and it is denoted as b here (Lin and Huang 2008). From (3.2) we have

$$\begin{aligned}
m_d(t) &= a(t) - a \cdot \exp\{-bW^*(t)\} - \exp\{-bW^*(t)\} \\
&\cdot \int_0^t a'(x) \cdot \exp\{bW^*(x)\}dx
\end{aligned} \tag{3.6}$$

The total number of faults $a(t)$ was usually assumed to be an exponential or linear function of time in the literature. Yamada et al. (1992) proposed two FDP models, with consideration of imperfect debugging, by assuming that the expected total number of faults increases exponentially and linearly with the testing time, respectively. An S-shaped concave FDP model was proposed in Pham et al. (1999) assuming that the total number of faults is a linear function of the testing time. In the following subsections, various TEF dependent FDP and FCP models are derived based on different assumptions on $a(t)$ and $f(x)$.

Paired Model 1

We assume that the total number of faults increases exponentially with the total testing effort consumed and the correction effort required is an exponential variable as

$$a(t) = a\exp\{\alpha W^*(t)\}, \alpha \geq 0 \tag{3.7}$$

$$f(x) = c\exp\{-cx\} \tag{3.8}$$

In this case we have

$$m_d(t) = \frac{ab}{b+\alpha}\left(\exp\{\alpha W^*(t)\} - \exp\{-bW^*(t)\}\right) \tag{3.9}$$

$$m_r(t) = \begin{cases} \dfrac{ab}{(b+\alpha)^2}(b\exp\{\alpha W^*(t)\} + \alpha\exp\{-bW^*(t)\}) \\ \quad -\dfrac{ab}{(b+\alpha)}(1+bW^*(t))\exp\{-bW^*(t)\}, c = b \\ \dfrac{a}{(1+\alpha/b)}\left(\dfrac{c\exp\{\alpha W^*(t)\} + \alpha\exp\{-cW^*(t)\}}{c+\alpha}\right. \\ \quad \left.+\dfrac{c\exp\{-bW^*(t)\} - b\exp\{-cW^*(t)\}}{b-c}\right), c \neq b \end{cases} \tag{3.10}$$

Actually (3.9) can be obtained by combining (3.6) and (3.7). (3.10) can be obtained by substituting (3.9) into (3.3) and (3.4). When $W^*(t)=t$, (3.9) is the same as the FDP model obtained in Yamada et al. (1992) for the case when the total number of faults is an exponential function of testing time. When $\alpha = 0$ and $W^*(t)=t$, (3.9) and (3.10) are the same as the paired model obtained in Wu et al. (2008) for the case of exponential debugging delay.

Paired Model 2

We assume that the total number of faults increases exponentially with the total testing effort consumed as given in (3.7) and the correction effort required is a gamma variable as

$$f(x) = \frac{\mu \exp\{-\mu x\}(\mu x)^{c-1}}{\Gamma(c)}, c, \mu > 0 \tag{3.11}$$

where $\Gamma(c) = \int_0^\infty \exp\{-y\}y^{c-1}\mathrm{d}y$ is the Euler gamma function.

Similarly we have

$$m_d(t) = \frac{ab}{b+\alpha}(\exp\{\alpha W^*(t)\} - \exp\{-bW^*(t)\}) \tag{3.12}$$

$$m_r(t) = \begin{cases} \dfrac{a\exp\{\alpha W^*(t)\}\Gamma(c,0,(b+\alpha)W^*(t))}{\left(1+\frac{\alpha}{b}\right)^{c+1}\Gamma(c)} \\ \quad -\dfrac{a\Gamma(c,0,bW^*(t))}{\left(1+\frac{\alpha}{b}\right)\Gamma(c)} + \dfrac{a\Gamma(c+1,0,bW^*(t))}{\left(1+\frac{\alpha}{b}\right)c\Gamma(c)}\bigg), \mu = b \\ \dfrac{a\exp\{\alpha W^*(t)\}\Gamma(c,0,(\mu+\alpha)W^*(t))}{\left(1+\frac{\alpha}{b}\right)\left(1+\frac{\alpha}{\mu}\right)^c\Gamma(c)} \\ \quad -\dfrac{a\exp\{-bW^*(t)\}\Gamma(c,0,(\mu-b)W^*(t))}{\left(1+\frac{\alpha}{b}\right)\left(1-\frac{b}{\mu}\right)^c\Gamma(c)}, \mu \neq b \end{cases} \tag{3.13}$$

where $\Gamma(\varepsilon_1,\varepsilon_2,\varepsilon_3) = \int_{\varepsilon_2}^{\varepsilon_3} e^{-y}y^{\varepsilon_1-1}\mathrm{d}y$ is a generalized incomplete gamma function.

When $\alpha = 0$ and $W^*(t)=t$, (3.12) and (3.13) are the same as the paired model obtained in Wu et al. (2008) for the case of gamma debugging delay.

Paired Model 3

We assume that the total number of faults increases linearly with the total testing effort consumed and the correction effort required is an exponential variable as

$$a(t) = a + sW^*(t), s \geq 0 \tag{3.14}$$

$$f(x) = c \exp\{-cx\} \tag{3.15}$$

In this case we have

$$m_d(t) = \left(a - \frac{s}{b}\right)(1 - \exp\{-bW^*(t)\}) + sW^*(t) \tag{3.16}$$

$$
m_r(t) =
\begin{cases}
\left(a - \dfrac{2s}{b}\right)(1 - (1 + bW^*(t))\exp\{-bW^*(t)\} \\
\quad + sW^*(t)(1 - \exp\{-bW^*(t)\}), c = b \\
\left(a - \dfrac{s}{b}\right)\left(1 + \dfrac{b\exp\{-cW^*(t)\} - c\exp\{-bW^*(t)\}}{c - b}\right) \\
\quad + sW^*(t) - \dfrac{s}{c}(1 - \exp\{-cW^*(t)\}), c \neq b
\end{cases}
\tag{3.17}
$$

Actually (3.16) can be obtained by combining (3.6) and (3.14). (3.17) can be obtained by substituting (3.16) into (3.3) and (3.4). When $W^*(t)=t$, (3.16) is the same as the FDP model obtained in Yamada et al. (1992) for the case when the total number of faults is a linear function of testing time. When $s = 0$ and $W^*(t)=t$, (3.16) and (3.17) are the same as the paired model obtained in Wu et al. (2008) for the case of exponential debugging delay.

Paired Model 4

We assume that the total number of faults increases linearly with the total testing effort consumed as given in (3.14) and the correction effort required is a gamma variable as given in (3.11).

Similarly we have

$$m_d(t) = \left(a - \frac{s}{b}\right)(1 - \exp\{-bW^*(t)\} + sW^*(t)) \tag{3.18}$$

$$m_r(t) = \begin{cases} \dfrac{sw(t)}{b\Gamma(c)}(bW^*(t)\Gamma(c,0,bW^*(t)) - \Gamma(c+1,0,bW^*(t))) \\[2ex] +\dfrac{(a-s/b)}{c\Gamma(c)}\Gamma(c+1,0,bW^*(t)), \mu = b \\[2ex] \dfrac{s}{\mu\Gamma(c)}(\mu W^*(t)\Gamma(c,0,\mu W^*(t)) - \Gamma(c+1,0,\mu W^*(t))+ \\[2ex] \dfrac{(a-s/b)}{\Gamma(c)}\Gamma(c,0,\mu W^*(t)) \\[2ex] -\dfrac{(a-s/b)\exp\{-bW^*(t)\}}{\Gamma(c)(1-b/\mu)^c}\Gamma(c,0,(\mu-b)W^*(t)), \mu \neq b \end{cases} \quad (3.19)$$

When $s = 0$ and $W^*(t)=t$, (3.18) and (3.19) are the same as the paired model obtained in Wu et al. (2008) for the case of gamma debugging delay.

3.2.4   A Summary of Various Testing-Effort Functions

Testing-effort functions that have been commonly used include constant, exponential, Rayleigh, Weibull, and logistic curves. Exponential curve and Rayleigh curve can be regarded as special cases of Weibull curve. The details are shown below.

Constant TEF

We assume that $w(t)$ is a constant. It can be expressed as

$$w(t) = w \quad (3.20)$$

Thus the cumulative testing effort $W(t)$ can be obtained as

$$W(t) = wt \quad (3.21)$$

It can be seen that the total testing effort consumed tends to positive infinity, when t approaches positive infinity. In the case that TEF is not considered, it can be regarded as considering $w(t) = 1$.

Weibull TEF

Weibull TEF is very flexible, and it can well fit most data that are often used in the study of SRGM. The cumulative TEF $W(t)$ is given by

$$W(t) = N(1 - \exp\{-\beta t^m\}) \tag{3.22}$$

where N is the expected total amount of testing effort that is required by software testing. β and m are the scale parameter and shape parameter, respectively. It should also be noted that the cumulative testing effort consumed is finite and tends to N when t approaches positive infinity.

Differentiating (3.22) gives

$$w(t) = N\beta m t^{m-1} \exp\{-\beta t^m\} \tag{3.23}$$

The exponential TEF is a special case of Weibull TEF when $m=1$. Exponential curve is suitable to describe the testing environment which has a monotonically declining testing-effort rate.

The Rayleigh TEF is a special case of Weibull TEF when $m=2$. Rayleigh testing-effort rate first increases to its peak and then decreases with a decelerating speed to zero asymptotically without reaching zero.

Logistic TEF

Logistic curve was first proposed in Parr (1980) as an alternative of Rayleigh curve. It exhibits similar behavior as Rayleigh curve, except during the initial stage of the project. The logistic cumulative TEF $W(t)$ is given by

$$W(t) = \frac{N}{1 + A \exp\{-\eta t\}} \tag{3.24}$$

where A is a constant parameter and η is the consumption rate of testing-effort expenditure. Similar to the Weibull case, the cumulative testing effort consumed is finite and tends to N when t approaches positive infinity.

Taking derivatives on both sides of (3.24) gives

$$w(t) = \frac{NA\eta}{\left(\exp\{\frac{\eta t}{2}\} + A \exp\{-\frac{\eta t}{2}\}\right)^2} \tag{3.25}$$

$w(t)$ reaches its maximum value when $t = \frac{\ln A}{\eta}$.

3.3 Parameter Estimation

Parameters in the different types of TEF are estimated by least square error (LSE). In order to select a TEF that best fits this dataset, some criteria are used to compare the performances of different TEFs.

1. RMSE

The root mean squared error (RMSE) is defined as

$$RMSE = \sqrt{\frac{1}{n} \cdot \sum_{i=1}^{n} (w(t_i) - w_i)^2} \qquad (3.26)$$

A smaller RMSE indicates a smaller fitting error and better performance.

2. Bias

The bias is defined as the sum of the deviation of the estimated testing curve from the actual data, as shown below:

$$Bias = \frac{1}{n} \cdot \sum_{i=1}^{n} (w(t_i) - w_i) \qquad (3.27)$$

3. Variance

The variance is defined as (Huang and Kuo 2002)

$$Variance = \sqrt{\frac{1}{n} \cdot \sum_{i=1}^{n} (w(t_i) - w_i - Bias)^2} \qquad (3.28)$$

4. RMSPE

The root mean squared prediction error (RMSPE) is defined as (Huang and Kuo 2002)

$$RMSPE = \sqrt{Variance + Bias^2} \qquad (3.29)$$

RMSPE is also a measure to depict how close the model predicts the observation.

After the suitable TEF is selected, the TEF together with the estimated parameters can be substituted into the software fault detection and correction models. The other parameters in the models can be obtained by fitting the models with the real numbers of detected faults and corrected faults observed in practice in order to minimize the RMSE.

3.4   Illustrative Example

3.4.1   Dataset Description

The dataset we use is from the System T1 data of the Rome Air Development Center (RADC) (Musa et al. 1987). Although this is quite an old dataset, it is widely used, and it contains both fault detection data and fault correction data. Additionally, it contains information of testing effort, which is characterized by computer time (CPU hours) consumed in each week. Hence this familiar dataset is used for illustration.

The cumulative numbers of detected faults and corrected faults during the first 21 weeks are shown in Table 3.1. During the time span, totally 300.1 CPU hours were consumed. Till the end of the testing, 136 faults were detected, and all of them were corrected.

Table 3.1   The dataset – System T1 (Peng et al. 2014)

Weeks	Computer time (CPU hours)	Cumulative number of detected faults	Cumulative number of corrected faults
1	4	2	1
2	4.3	2	2
3	2	2	2
4	0.6	3	3
5	2.3	4	4
6	1.6	6	4
7	1.8	7	5
8	14.7	16	7
9	25.1	29	13
10	4.5	31	17
11	9.5	42	18
12	8.5	44	32
13	29.5	55	37
14	22	69	56
15	39.5	87	75
16	26	99	85
17	25	111	97
18	31.4	126	117
19	30	132	129
20	12.8	135	131
21	5	136	136

3.4.2 Select the Most Suitable TEF for This Dataset

Estimated parameters and comparison results for different TEFs are shown in Table 3.2. Figure 3.1 is plotted for graphical illustration.

It can be seen that logistic TEF has the smallest RMSE, variance, and RMSPE and also has a smaller bias than Weibull TEF. Figure 3.1 also shows that logistic TEF fits best. Thus logistic TEF is adopted for further analysis.

3.4.3 Performance Analysis

The paired model (3.9) and (3.10) is used for illustration. After substituting the cumulative logistic testing-effort function (3.24) with the estimated parameters

Table 3.2 Estimated parameters and comparison results for different TEFs (Peng et al. 2014)

TEF	Estimated parameters	RMSE	Bias	Variance	RMSPE
Constant	$w = 14.29$	12.11	−0.00048	12.11	12.11
Weibull	$N = 407.0830$	7.86	1.2615	7.76	7.86
	$\beta = 2.064\text{E-}4$				
	$m = 2.923$				
Logistic	$N = 321.482$	6.7828	−0.5778	6.7570	6.7818
	$\eta = 0.3826$				
	$A = 423.788$				

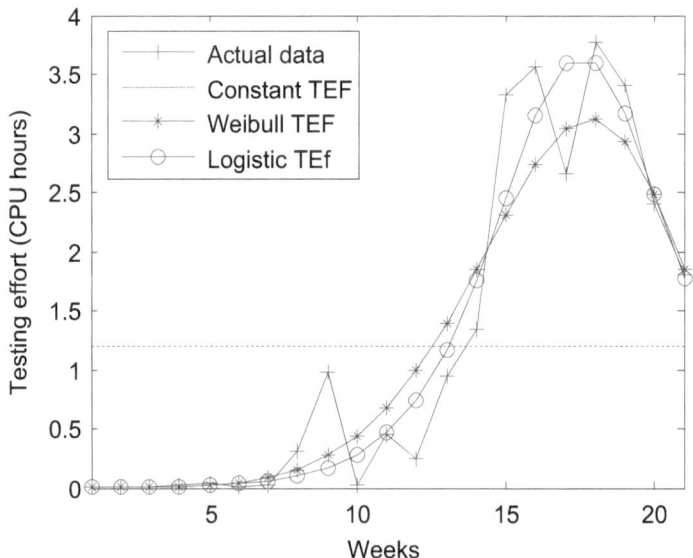

Fig. 3.1 Observed/estimated TEF for the real dataset (Peng et al. 2014)

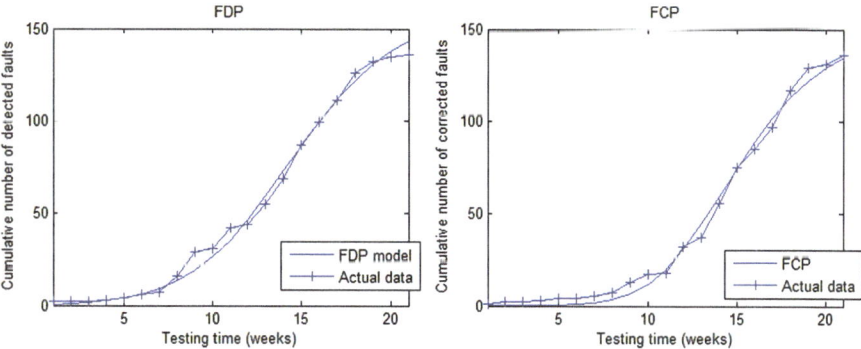

Fig. 3.2 Paired model fitted against the real dataset (Peng et al. 2014)

$N = 321.482$, $\eta = 0.3826$, and $A = 423.788$ into (3.9) and (3.10), the paired model is applied to fit against the real dataset. The LSE estimation of the parameters are obtained as $a = 100.97$, $b = 0.0094$, $\alpha = 0.0021$, and $c = 0.0418$. According to the estimated parameters, there are about 100.97 faults at the beginning of testing. The total number of faults when t approaches infinity is expected to be $\lim_{t \to \infty} a$ $(t) = 198.01$. Figure 3.2 is plotted for graphical illustration.

3.5   Software Release Policies

Determination of the optimal release time is a critical decision for software projects and has been studied in many papers (Boland and Chuí 2007; Inoue and Yamada 2007; Li et al. 2010). As cost and reliability requirements are of great concern, they are often used to determine the time to stop the testing and release the software (Lin and Huang 2008; Peng et al. 2014, 2015).

3.5.1   Software Release Policy Based on Reliability Criterion

Software reliability is defined as the probability that no failure occurs during time interval (T, T+ΔT] given that the software is released at time T. Considering that software normally doesn't change in operational phase, the reliability function is

$$R(\Delta T | \mathrm{T}) = \exp\{-\lambda_d(T) \cdot \Delta T\} \qquad (3.30)$$

If R_1 is the reliability target and T_{LC} is the length of the software life cycle, the time when the reliability of the software reaches R_1 can be obtained as $T_1 = \inf \{\lambda_d$ $(T) \leq \ln (1/R_1)/\Delta T : T \in [0, T_{\mathrm{LC}}]\}$.

3.5.2 Software Release Policy Based on Cost Criterion

Beside the reliability requirement, we can also discuss the optimal release time based on the total cost during the software testing phase and the operational phase. With the incorporation of FCP $m_r(t)$, the cost model can be expressed as

$$C(T) = c_1 m(T) + c_2(m_d(T_{LC}) - m_r(T)) + c_3 W^*(T) \qquad (3.31)$$

where c_1 is the cost of fixing a fault during the testing phase, c2 is the cost of fixing a fault during the operational phase ($c_2 > c_1 > 0$), and c_3 is the unit cost for testing effort consumed during testing. By minimizing the cost model with respect to T, the optimal release time T_c can be obtained.

Differentiating both sides of (3.31), we have

$$C'(T) = c_3 w(T) - (c_2 - c_1)\lambda_r(T) \qquad (3.32)$$

Furthermore we have $C'(0) = c_3 w(0) > 0$. Let $z_1 \le z_2 \le \ldots \le z_n$ be all the solutions to $\lambda_r(T)/w(T) = c_3/(c_2 - c_1)$ during $(0, T_{LC})$. If $n = 2k$ $(k \ge 0)$, T_c can be determined as $T_c = \mathrm{argmin}_{T=0, z_2, \ldots, z_{2k}} C(T)$. Otherwise $n = 2k + 1$ and T_c can be determined as $T_c = \mathrm{argmin}_{T=0, z_2, \ldots, z_{2k}, T_{LC}} C(T)$.

3.5.3 Software Release Policy Based on Mixed Criterion

When both reliability requirements and the total cost are considered, our goal is to determine the optimal release time T^* which minimizes the total cost without compromising the reliability requirements. Thus the problem can be formulated as

Minimize $C(T) = c_1 m_r(T) + c_2(m_d(T_{LC}) - m_r(T)) + c_3 W^*(T)$
Subject to $R(\Delta T \mid T) = \exp[-\lambda_d(T)\Delta T] \ge R_1$

The time axis (T_1, T_{LC}) can be divided into four types of intervals such that both $R(\Delta T|T)$ and $C(T)$ increase on type 1 intervals, both $R(\Delta T|T)$ and $C(T)$ decrease on type 2 intervals, and $R(\Delta T|T)$ increases, while $C(T)$ decreases on type 3 intervals, and $R(\Delta T|T)$ decreases, while $C(T)$ increases on type 4 intervals. The candidates for T^* comprise of the minimum T in each type 1 interval that satisfies $R(\Delta T|T) \ge R_1$, the maximum T in each type 2 interval that satisfies $R(\Delta T|T) \ge R_1$, the end points of type 3 intervals which satisfy $R(\Delta T|T) \ge R_1$, and the initial points of type 4 intervals which satisfy $R(\Delta T|T) \ge R_1$. T^* equals to the candidate which incurs the lowest cost.

3.5.4 Numerical Examples for Software Release Policy

For illustration, we consider the first paired model (3.9) and (3.10) with parameters estimated as $a = 100.97$, $b = 0.0094$, $\alpha = 0.0021$, and $c = 0.0418$ and logistic TEF with parameters estimated as $N = 321.482$, $\eta = 0.3826$, and $A = 423.788$. We also assume $T_{LC} = 300$, $c_1 = \$300$, $c_2 = \$2000$, $c_3 = \$700$, $\Delta T = 10$, and $R_1 = 0.95$.

From (3.9) we have

$$
\begin{aligned}
\lambda_d(T) &= b \cdot w(T) \cdot \left(\frac{a\alpha}{b+\alpha} \exp\{\alpha W^*(T)\} + \frac{ab}{b+\alpha} \exp\{-bW^*(T)\} \right)' \\
&= \frac{9033.4\exp\{0.0021W^*(T)\} + 40439\exp\{-0.0094W^*(T)\}}{(\exp\{0.1913T\} + 423.788\exp\{-0.1913T\})^2}.
\end{aligned}
$$

It can be seen that $\lambda_d(T)$ increases from on [0, 14.112] and decreases on (14.112, 300). Solving $\lambda_d(T) = \ln(1/R_1)/\Delta T = 0.0051$ gives $T_1 = 39.626$. The reliability requirement is satisfied if the software is released after 39.626 weeks of testing.

From (3.46) and (3.47), we have

$$
\begin{aligned}
C(T) &= c_1 m_r(T) + c_2(m_d(T_{LC}) - m_r(T)) + c_3 W^*(T) \\
&= 253590 - 1700 m_r(T) + 700 W^*(T)
\end{aligned}
\tag{3.33}
$$

$$
C'(T) = 700w(T) - 1700\lambda_r(T)
\tag{3.34}
$$

In addition we have

$$
\begin{aligned}
\frac{\lambda_r(T)}{w(T)} &= \frac{a\alpha bc}{(b+\alpha)(c+\alpha)} (\exp\{\alpha W^*(T)\} - \exp\{-cW^*(T)\}) \\
&+ \frac{ab^2 c(\exp\{-cW^*(T)\} - \exp\{-bW^*(T)\})}{(b+\alpha)(b-c)} \\
&= 0.165^* \exp\{0.0021W^*(T)\} - 1.1659\exp\{-0.0418W^*(T)\} \\
&+ 1.0009\exp\{-0.0094W^*(T)\}
\end{aligned}
$$

Solving $\frac{\lambda_r(T)}{w(T)} = \frac{700}{1700}$ gives $T = 8.013$ and 16.848. Thus $C(T)$ increases on [0, 8.013], decreases on (8.013, 16.848), and increases on [16.848, 300]. The optimal release time which minimizes the total cost is $T_c = 16.848$. The corresponding total cost is $C(T) = \$217820$.

As both $R(\Delta T | T)$ and $C(T)$ increase on $[T_1, 300]$, the optimal software release time $T^* = T_1 = 39.626$. Figure 3.3 is plotted for graphical illustration.

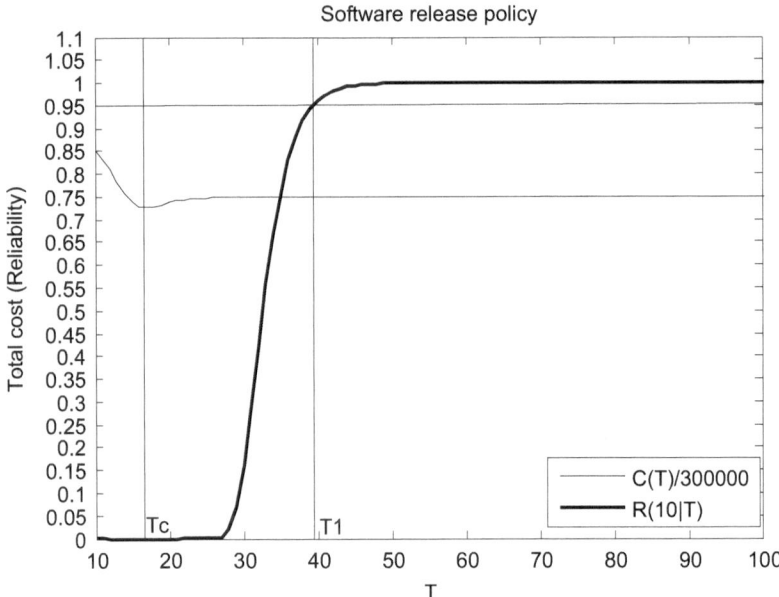

Fig. 3.3 Plot of total cost function and software reliability function (Peng et al. 2014)

3.6 Summary

This chapter studies testing-effort function dependent software FDP and FCP with incorporation of imperfect debugging. Testing resource is usually not constantly allocated during software testing phase, which can largely influence the fault detection rate and the time needed to correct the detected faults. For example, the debugger may spend a week without doing any testing work and work very hard in the following few days. In addition, it is natural for debuggers to make mistakes and introduce new faults during testing. The debuggers tend to introduce more faults if more testing effort is consumed since the code has experienced more changes. In order to capture the influences of testing resource allocation and fault introduction on both FDP and FCP, we first derive FDP incorporating testing-effort function and the fault introduction effect and then obtain FCP as delayed FDP with a correction effort. Various paired FDP and FCP models are obtained based on different assumptions on fault introduction and correction effort. It can be seen that our model is quite general and flexible. Some simpler models are the special cases of our models. An example is presented to illustrate the application of the paired models. The optimal release policy under different criteria is also studied.

References

Boland, P. J., & Chuí, N. N. (2007). Optimal times for software release when repair is imperfect. *Statistics & Probability Letters, 77*, 1176–1184.

Cai, K. Y., Cao, P., Dong, Z., & Liu, K. (2010). Mathematical modeling of software reliability testing with imperfect debugging. *Computers & Mathematics with Applications, 59*(10), 3245–3285.

Chang, Y. C., & Liu, C. T. (2009). A generalized JM model with applications to imperfect debugging in software reliability. *Applied Mathematical Modeling, 33*(9), 3578–3588.

Chiu, K. C., Huang, Y. S., & Lee, T. Z. (2008). A study of software reliability growth from the perspective of learning effects. *Reliability Engineering & System Safety, 93*(10), 1410–1421.

Demarko, T. (1982). *Controlling software projects: Management, measurement and estimation.* Englewood Cliffs: Prentice-Hall.

Gokhale, S. S., Lyu, M. R., & Trivedi, K. S. (2006). Incorporating fault debugging activities into software reliability models: A simulation approach. *IEEE Transactions on Reliability, 55*(2), 281–292.

Hu, Q. P., Xie, M., Ng, S. H., & Levitin, G. (2007). Robust recurrent neural network modeling for software fault detection and correction prediction. *Reliability Engineering & System Safety, 92*(3), 332–340.

Huang, C. Y. (2005). Performance analysis of software reliability growth models with testing-effort and change-point. *Journal of Systems and Software, 76*(2), 181–194.

Huang, C. Y., & Kuo, S. Y. (2002). Analysis and assessment of incorporating logistic testing effort function into software reliability modeling. *IEEE Transactions on Reliability, 51*(3), 261–270.

Hwang, S., & Pham, H. (2009). Quasi-renewal time-delay fault-removal consideration in software reliability modeling. *IEEE Transactions on Systems Man and Cybernetics Part A-Systems and Humans, 39*(1), 200–209.

Inoue, S., & Yamada, S. (2007). Generalized discrete software reliability modeling with effect of program size. *IEEE Transactions on Systems, Man, and Cybernetics: Part A – Systems and Humans, 37*(2), 170–179.

Jain, M., & Gupta, R. (2011). Optimal release policy of module-based software. *Quality Technology and Quantitative Management, 8*(2), 147–165.

Jha, P. C., Gupta, D., Yang, B., & Kapur, P. K. (2009). Optimal testing resource allocation during module testing considering cost, testing effort and reliability. *Computers & Industrial Engineering, 57*(3), 1122–1130.

Jia, L. X., Yang, B., Guo, S. C., & Park, D. H. (2010). Software reliability modeling considering fault correction process. *IEICE Transactions on Information and Systems, E93D*(1), 185–188.

Kang, H. G., Lim, H. G., Lee, H. J., Kim, M. C., & Jang, S. C. (2009). Input-profile-based software failure probability quantification for safety signal generation systems. *Reliability Engineering & System Safety, 94*(10), 1542–1546.

Kapur, P. K., Goswami, D. N., Bardhan, A., & Singh, O. (2008). Flexible software reliability growth model with testing effort dependent learning process. *Applied Mathematical Modelling, 32*, 1298–1307.

Kapur, P. K., Shatnawi, O., Aggarwal, A., & Kumar, R. (2009). Unified framework for developing testing effort dependent software reliability growth models. *WSEAS Transactions on Systems, 8*(4), 521–531.

Kapur, P. K., Pham, H., Anand, S., & Yadav, K. (2011). A unified approach for developing software reliability growth models in the presence of imperfect debugging and error generation. *IEEE Transactions on Reliability, 60*(1), 331–340.

Kim, H. S., Park, D. H., & Yamada, S. (2009). Bayesian optimal release time based on inflection S-shaped software reliability growth model. *IEICE Transactions on Fundamentals of Electronics, Communications and Computer Sciences, E92A*(6), 1485–1493.

Li, X., Xie, M., & Ng, S. H. (2010). Sensitivity analysis of release time of software reliability models incorporating testing effort with multiple change-points. *Applied Mathematical Modelling, 34*(11), 3560–3570.

Lin, C. T., & Huang, C. Y. (2008). Enhancing and measuring the predictive capabilities of testing-effort dependent software reliability models. *Journal of Systems and Software, 81*, 1025–1038.

Musa, J. D., Iannino, A., & Okumono, K. (1987). *Software reliability, measurement, prediction and application.* New York: McGraw Hill.

Okamura, H., Dohi, T., & Osaki, S. (2013). Software reliability growth models with normal failure time distributions. *Reliability Engineering & System Safety, 116*, 135–141.

Parr, F. N. (1980). An alternative to the Rayleigh curve for software development effort. *IEEE Transactions on Software Engineering, 6*(3), 291–296.

Peng, R., Li, Y. F., Zhang, W. J., & Hu, Q. P. (2014). Testing effort dependent software reliability model for imperfect debugging process considering both detection and correction. *Reliability Engineering and System Safety, 126*, 37–43.

Peng, R., Li, Y. F., Zhang, J. G., & Li, X. (2015). A risk-reduction approach for optimal software release time determination with the delay incurred cost. *International Journal of Systems Science, 46*(9), 1628–1637.

Pham, H., Nordmann, L., & Zhang, X. (1999). A general imperfect software debugging model with s-shaped fault detection rate. *IEEE Transactions on Reliability, 48*, 169–175.

Pievatolo, A., Ruggeri, F., & Soyer, R. (2012). A Bayesian hidden Markov model for imperfect debugging. *Reliability Engineering & System Safety, 103*, 11–21.

Schneidewind, N. F. (1975). *Analysis of error processes in computer software. Proceedings of the International Conference on Reliable Software* (pp. 337–346). Los Alamitos: IEEE Computer Society Press.

Shyur, H. J. (2003). A stochastic software reliability model with imperfect-debugging and change-point. *Journal of Systems and Software, 66*, 135–141.

Stikkel, G. (2006). Dynamic model for the system testing process. *Information and Software Technology, 48*, 578–585.

Wu, Y. P., Hu, Q. P., Xie, M., & Ng, S. H. (2008). Modeling and analysis of software fault detection and correction process by considering time dependency. *IEEE Transactions on Reliability, 56*(4), 629–642.

Xie, M., & Yang, B. (2003). A study of the effect of imperfect debugging on software development cost. *IEEE Transactions on Software Engineering, 29*(5), 471–473.

Xie, M., Hu, Q. P., Wu, Y. P., & Ng, S. H. (2007). A study of the modeling and analysis of software fault-detection and fault-correction processes. *Quality and Reliability Engineering International, 23*, 459–470.

Yamada, S., Tokuno, K., & Osaki, S. (1992). Imperfect debugging models with fault introduction rate for software reliability assessment. *International Journal of Systems Science, 23*(12), 2241–2252.

Yang, B., Li, X., Xie, M., & Tan, F. (2010). A generic data-driven software reliability model with model mining technique. *Reliability Engineering & System Safety, 95*(6), 671–678.

Zhang, X. M., Teng, X. L., & Pham, H. (2003). Considering fault removal efficiency in software reliability assessment. *IEEE Transactions on Systems, Man and Cybernetics: Part A – Systems and Humans, 33*, 114–120.

Chapter 4
Software Reliability Models Considering Fault Dependency

This chapter incorporates the fault dependency into the fault detection process and fault correction process. In particular, faults are classified as leading faults and dependent faults, where dependent faults become detectable only after the corresponding leading faults are corrected. Different paired models of the fault detection process and the fault correction process are constructed based on different assumptions of the debugging delay. The applications of the proposed models are illustrated with real datasets, and the optimal release time policy is studied.

4.1 Introduction

During the past three decades, numerous software reliability growth models (SRGMs) have been proposed (Lyu 1996; Gutjahr 2001; Chang and Liu 2009; Shatnawi 2009; Lin and Li 2014; Wang et al. 2016). Among these models, nonhomogeneous Poisson process (NHPP) models are the most commonly accepted (Lee et al. 2004; Tamura and Yamada 2006; Zhao et al. 2006; Shatnawi 2014; Peng et al. 2015). Although NHPP models are more mathematically tractable, they are developed under some strong assumptions on the software testing process. Specifically, NHPP models assume immediate fault removal and fault independency. To adapt to different practical software testing environments, generalizations of traditional models by relaxing the assumptions have been proposed (Gokhale et al. 2004; Huang and Huang 2008; Kapur et al. 2008; Lin and Huang 2009; Okamura and Dohi 2011; Okamura et al. 2013).

In practical software testing, each detected fault has to be reported, diagnosed, removed, and verified before it can be noted as corrected. Consequently, the time spent for fault correction activity is not negligible. In fact, this debugging lag can be an important element in making decisions (Zhang et al. 2003; Jia et al. 2010). Therefore, it is necessary to incorporate the debugging lag into the modeling

R. Peng et al., *Software Fault Detection and Correction: Modeling and Applications*, SpringerBriefs in Computer Science, https://doi.org/10.1007/978-981-13-1162-8_4

framework, i.e., to model both the fault detection process (FDP) and fault correction process (FCP). The idea of modeling FCP was first proposed in Schneidewind (1975), where a constant lag was used to model the FCP after fault detection. Clearly, the constant correction time assumption is restrictive for various types of faults and different correction profiles. For instance, data collected from practical testing projects show that the correction time can be fitted by the exponential and log-normal distributions (Musa et al. 1987). In addition, the correction time may show a growing trend during the whole testing cycle, as later detected faults can be more difficult to correct. Some extensions were made in Lo and Huang (2006) and Xie et al. (2007) by incorporating other assumptions of debugging delay. Hu et al. (2007) studied a data-driven artificial neural network model for the prediction of FDP and FCP. Shibata et al. (2007) used the fault detection/correction profile to quantify the maintainability of the software. Some paired FDP and FCP models were proposed in Peng et al. (2014), where testing effort function and fault introduction were included.

Traditional NHPP models assume the statistical independency between successive software failures. Actually, it can hardly be true in practice, as some faults are not detectable until some other fault has been corrected because of logical dependency. Moreover, the common practice of mixing testing strategies can lead to the dependency of failures (Goseva-Popstojanova and Trivedi 2000). With a failure detected, there is a higher chance for another related failure or a cluster of failures to occur in the near future. From this point of view, faults can be classified into mutually independent and dependent types with respect to path-based testing approach. This issue was addressed in Kapur and Younes (1995), where an extended NHPP SRGM was proposed. Following the basic assumptions on fault dependency from Kapur and Younes (1995), some more recent works have been developed to incorporate fault dependency into paired FDP and FCP models (Huang and Lin 2006). The statistical inference of the software reliability model with fault dependency has been discussed by Yang et al. (2008). However, most of the studies only focusing on the FDP model parameters are estimated based on the FDP data only, whereas the collected information from FCP is neglected.

To remedy the deficiency, we incorporate the fault dependency into the paired FDP and FCP models. Instead of assuming a single type of fault, this study classifies the faults in the testing process into leading faults and dependent faults. The leading faults occur independently following an NHPP, while the dependent faults are only detectable after the related leading faults being corrected. Different from (Huang and Lin 2006) which modeled the FDP and the FCP as a single, synthesized fault detection and correction process, we modeled the FDP and FCP for the leading faults and the dependent faults separately. Subsequently, the FDP and FCP model for the aggregated, observable faults can be readily obtained. With different formulation of debug delays, we can derive various FDP and FCP models. Hence, the proposed models admit a wide applicability that can account for different software reliability growth schemes.

The rest of this chapter is organized as follows. Section 4.2 formulates the general modeling framework of paired FDP and FCP with the incorporation of fault dependency. Besides, several paired FDP and FCP models are derived based on different

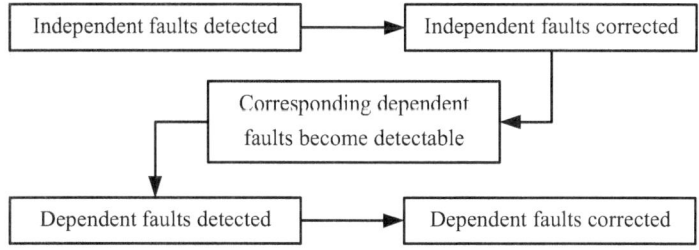

Fig. 4.1 Relationship of leading faults and dependent faults (Peng and Zhai 2017)

assumptions on the debugging lag. Parameter estimation method is introduced in Sect. 4.3. Two real datasets are used to illustrate the application of the proposed models in Sect. 4.4. Section 4.4.1 derives the optimal software release policy under the proposed framework.

4.2   The Modeling Framework

The practical fault-oriented software testing process can be described as FDP and FCP. During the testing process, faults can only be corrected after being detected. Furthermore, faults embedded in the original software system can be categorized into leading and dependent faults. Those faults that can be detected and corrected independently are defined as leading faults or independent faults. Other faults that remain undetectable until their corresponding leading faults are removed are defined as dependent faults. Figure 4.1 displays the relationship between leading faults and dependent faults.

Suppose the detection and correction of leading faults are independent. Then, the FDP for leading faults (FDP_L) and the FCP for leading faults (FCP_L) can be modeled similarly as in Xie et al. (2007). Because dependent faults can only be detectable after the corresponding leading faults have been corrected, the FDP for dependent faults (FDP_D) can be modeled as a delayed process of FCP_L. The FCP for dependent faults (FCP_D) can be modeled as a delayed process of FDP_D. The modeling framework is characterized by the mean value function for each sub-process.

4.2.1   Modeling FDP_L

We assume that FDP_L follows a NHPP, and the expected number of leading fault detected during $(t, t + \Delta t]$ is proportional to the number of undetected leading faults at time t. Thus, we have

$$\frac{dm_{d1}(t)}{dt} = b(t)(a_1 - m_{d1}(t)), \tag{4.1}$$

where $b(t)$ is the fault detection rate at time t and a_1 is number of leading faults at the beginning. With the initial condition that $m_{d1}(t) = 0$, it can be derived from (4.1) that

$$m_{d1}(t) = a_1 \left(1 - \exp\left\{ -\int_0^t b(s)ds \right\} \right). \tag{4.2}$$

Different $m_{d1}(t)$ can be obtained based on different $b(t)$. Specially, when $b(t)$ is a constant, we have

$$m_{d1}(t) = a_1 \left(1 - e^{-bt} \right), \tag{4.3}$$

which is the GO model (Goel and Okumoto 1979). When $b(t) = \frac{b^2 t}{1+bt}$, we have

$$m_{d1}(t) = a_1 \left(1 - (1 + bt)e^{-bt} \right), \tag{4.4}$$

which has the same form as the Yamada delayed S-shaped model (Yamada et al. 1984).

4.2.2 Modeling FCP$_L$

FCP$_L$ can be regarded as a delayed process of FDP$_L$, and different models can be used to accommodate the debugging delay. Xie et al. (2007) pointed out that debugging lags could be assumed constant, time dependent, or random. If the debugging lag is not random, the FCP$_L$ can be derived from FDP$_L$ as $m_{r1}(t) = m_{d1}(t - \delta(t))$. Otherwise, we have $m_{r1}(t) = E_{\delta(t)}[m_{d1}(t - \delta(t))]$ if $\delta(t)$ is a random variable. In particular, if the debugging lag is assumed to be an exponentially distributed random variable, i.e., $\delta(t){\sim}\text{Exp}(c)$, we have

$$m_{r1}(t) = c \int_0^t m_{d1}(t - s)\exp\{-cs\}ds. \tag{4.5}$$

Taking the derivatives of both sides with respect to t, we can obtain that

$$\lambda_{r1}(t) = c(m_{d1}(t) - m_{r1}(t)). \tag{4.6}$$

This indicates that the expected number of faults corrected during $(t, t + \Delta t]$ is proportional to the number of detected but uncorrected faults at time t. We call c as the fault correction rate.

4.2.3 Modeling FDP$_D$ and FCP$_D$

For these dependent faults, they can only be detected after the corresponding leading faults are removed. Hence, the proportion of the detectable dependent faults in the dependent faults is equal to the proportion of the corrected leading faults in the leading faults. Suppose the number of dependent faults is a_2. Then, the expected number of detectable dependent faults is $a_2 m_{r1}(t)/a_1$ up to time t. Furthermore, because leading faults and dependent faults are detected under the same testing environment, it is reasonable to assume that the fault detection rate for dependent faults is the same as the fault detection rate for leading faults. Therefore,

$$\frac{dm_{d2}(t)}{dt} = b(t)\left(\frac{a_2 m_{r1}(t)}{a_1} - m_{d2}(t)\right). \tag{4.7}$$

With the initial condition $m_{d2}(0) = 0$, we can derive from (4.7) based on $m_{r1}(t)$ and $b(t)$ that

$$m_{d2}(t) = \frac{a_2}{a_1} m_{r1}(t)$$
$$-\frac{a_2}{a_1}\exp\left\{-\int_0^t b(s)ds\right\}\int_0^t \lambda_{r1}(s)\exp\left\{\int_0^s b(u)du\right\}ds. \tag{4.8}$$

Particularly, when $b(t) = b$, we have

$$m_{d2}(t) = \frac{a_2}{a_1} m_{r1}(t) - \frac{a_2}{a_1}\exp\{-bt\}\int_0^t \lambda_{r1}(s)\exp\{bs\}ds. \tag{4.9}$$

After obtaining $m_{d2}(t)$, different $m_{d2}(t)$ can be derived based on $m_{d2}(t)$ and different assumptions of the debugging delay, as for FCP$_L$.

4.2.4 Combined Models

Since faults are exclusively classified as leading faults and dependent faults, we have

$$a = a_1 + a_2, \tag{4.10}$$
$$m_d(t) = m_{d1}(t) + m_{d2}(t), \tag{4.11}$$
$$m_r(t) = m_{r1}(t) + m_{r2}(t). \tag{4.12}$$

4.2.5 Specific Models for Dependent FDP and FCP

Fault detection rate function $b(t)$ is usually assumed to be constant, i.e., $b(t) = b$, see Huang et al. (2007) and Lin and Huang (2008). In this case, $m_{d1}(t) = ap(1 - e^{-bt})$ from (4.3), where p is the proportion of leading faults, $p = a_1/a$. Based on different assumptions of the relations between $m_{r1}(t)$ and $m_{d1}(t)$, different can be derived. Furthermore, we note that can be obtained according to (4.9) as long as $m_{r1}(t)$ is specified. In the following, we consider some practical cases for the debugging lag and solve the corresponding $m_d(t)$ and $m_r(t)$.

Constant Debugging Lag

It is assumed that the correction of each fault takes the same time, i.e., $\delta(t) = \delta$. Thus, $\lambda_{ri}(t) = \lambda_{di}(t - \delta)$, $i = 1, 2$. Consequently, we have

$$m_{r1}(t) = \begin{cases} 0, t < \delta \\ ap\left(1 - e^{-b(t-\delta)}\right), t \geq \delta \end{cases} \tag{4.13}$$

According to (4.9), we can readily obtain that

$$m_{d2}(t) = \begin{cases} 0, t < \delta \\ a(1-p)\left(1 - (1 + bt - b\delta)e^{-b(t-\delta)}\right), t \geq \delta \end{cases} \tag{4.14}$$

Then, $m_d(t)$ is solved as

$$m_d(t)$$
$$= \begin{cases} ap\left(1 - e^{-bt}\right), t < \delta \\ ap\left(1 - e^{-bt}\right) + a(1-p)\left(1 - (1 + bt - b\delta)e^{-b(t-\delta)}\right), t \geq \delta \end{cases} \tag{4.15}$$

According to the assumption of constant debugging delay, we have

$$m_r(t)$$
$$= \begin{cases} 0, t < \delta \\ ap\left(1 - e^{-b\,(t-\delta)}\right), \delta \leq t < 2\delta \\ ap\left(1 - e^{-b(t-\delta)}\right) + a(1-p)\left(1 - (1 + bt - 2b\delta)e^{-b(t-2\delta)}\right), t \geq 2\delta \end{cases} \tag{4.16}$$

Time-Dependent Debugging Lag

In practice, later discovered faults may be more difficult to correct. To model such a phenomenon, we assume $\delta(t) = \frac{\ln(1+\gamma t)}{b}$ and $m_{ri}(t) = m_{di}\left(t - \frac{\ln(1+\gamma t)}{b}\right)$, $i = 1, 2$, and where $0 < \gamma < b$ actually, under this assumption, we have

$$m_{r1}(t) = m_{d1}\left(t - \frac{\ln(1+\gamma t)}{b}\right) = ap\left(1 - (1+\gamma t)e^{-bt}\right), \qquad (4.17)$$

which is a general form of the delayed NHPP model; see Yamada et al. (1984).

Combining (4.9) and (4.17) together gives $m_{d2}(t)$, with which $m_d(t)$ can be obtained as

$$m_d(t) = a\left(1 - e^{-bt}\right) - a(1 - p)\left(bt + \frac{b\gamma t^2}{2}\right)e^{-bt}. \qquad (4.18)$$

Moreover, as $m_r(t) = m_d\left(t - \frac{\ln(1+\gamma t)}{b}\right)$, we have

$$\begin{aligned}
m_r(t) = &\ a\left(1 - (1+\gamma t)e^{-bt}\right) \\
&- a(1 - p)(1 + \gamma t)\left(bt - (1+\gamma t)\ln(1+\gamma t) + \frac{b\gamma t^2}{2}\right. \\
&\left. + \frac{\gamma \ln^2(1+\gamma t)}{2b}\right)e^{-bt}
\end{aligned} \qquad (4.19)$$

Exponentially Distributed Random Debugging Lag

As obtained in Sect. 4.1.B, in this case the number of faults corrected during time interval $(t, t + \Delta t]$ is proportional to the number of detected but uncorrected faults at time t. Based on (4.5), $m_{r1}(t)$ can be obtained as

$$m_{r1}(t) = \begin{cases} ap\left(1 - (1+bt)e^{-bt}\right), & c = b \\ ap\left(1 + \dfrac{be^{-ct} - ce^{-bt}}{c - b}\right), & c \neq b \end{cases} \qquad (4.20)$$

Then, $m_{d2}(t)$ can be derived based on $m_{r1}(t)$ according to (4.9). Subsequently, $m_d(t)$ is obtained as the summation of $m_{d1}(t)$ and $m_{d2}(t)$:

$m_d(t)$

$$= \begin{cases} a(1 - e^{-bt}) - a(1 - p)\left(bt + \dfrac{b^2 t^2}{2}\right)e^{-bt}, c = b \\ a(1 - e^{-bt}) - a(1 - p)\left(\dfrac{bcte^{-bt}}{c - b} + \dfrac{b^2\left(e^{-ct} - e^{-bt}\right)}{(c - b)^2}\right), c \neq b \end{cases} \quad (4.21)$$

From $m_{ri}(t) = c\displaystyle\int_0^t m_{di}(t - s)\exp\{-cs\}ds$, we note that $m_r(t) = c\displaystyle\int_0^t m_d(t - s)$ $\exp\{-cs\}ds$ also holds. Therefore, $m_r(t)$ is readily obtained as

$m_r(t)$

$$= \begin{cases} a(1 - (1 + bt)e^{-bt}) - a(1 - p)\left(\dfrac{b^2 t^2}{2} + \dfrac{b^3 t^3}{6}\right)e^{-bt}, c = b \\ a\left(1 + \dfrac{be^{-ct} - ce^{-bt}}{c - b}\right) - \dfrac{abc(1 - p)}{(c - b)^2}\left(\dfrac{(b + c)\left(e^{-ct} - e^{-bt}\right)}{c - b} + bte^{-ct} + cte^{-bt}\right), c \neq b \end{cases}$$

$$(4.22)$$

4.3   Parameter Estimation

In real applications, the parameters of the models proposed in this chapter need to be estimated. In particular, the numbers of detected faults and corrected faults should be recorded. After that, the models can be fitted against the real data to find out the combination of parameter values which minimizes the deviation between the theoretical models and the data. A typical way is to use the least squares method, which minimizes the mean squared error (MSE) between the estimated cumulative numbers of detected and corrected faults and the actual cumulative numbers of detected and corrected faults.

In the case of both fault detection process and fault correction process, the mean squared error is the average between the mean squared error for the fault detection process and that for the fault correction process. Therefore, it is calculated as

$$\text{MSE} = \frac{1}{2}(\text{MSE}_d + \text{MSE}_r)$$

$$= \frac{1}{2n}\sum_{i=1}^{n}\left[(m_d(t_i) - m_{d,i})^2 + (m_r(t_i) - m_{r,i})^2\right] \quad (4.23)$$

where $m_{d,i}$ and $m_{r,i}$ are the observed cumulative numbers of detected faults and corrected faults at time $t_i, i = 1, \ldots, n$.

4.4 Numerical Example

4.4.1 Description of the Datasets

The first dataset is from the System T1 data of the Rome Air Development Center (RADC) (Musa et al. 1987). This dataset is widely used, and it contains both fault detection data and fault correction data. The cumulative numbers of detected faults and corrected faults during the first 21 weeks are shown in Table 4.1. During the time span, 300.1 h of computer time were consumed, and 136 faults were removed. Computer time is indeed the time which is verily spent on software testing; thus, it is used here to characterize the testing progress.

The second dataset we use is from the testing process on a middle-size software project (Xie et al. 2007; Wu et al. 2008). The cumulative numbers of detected faults and corrected faults during the first 17 weeks are listed in Table 4.2.

Table 4.1 The dataset: System T1 (Peng and Zhai 2017)

Weeks	Computer time (CPU hours)	Cumulative number of detected faults (m_d)	Cumulative number of corrected faults (m_r)
1	4	2	1
2	8.3	2	2
3	10.3	2	2
4	10.9	3	3
5	13.2	4	4
6	14.8	6	4
7	16.6	7	5
8	31.3	16	7
9	56.4	29	13
10	60.9	31	17
11	70.4	42	18
12	78.9	44	32
13	108.4	55	37
14	130.4	69	56
15	169.9	87	75
16	195.9	99	85
17	220.9	111	97
18	252.3	126	117
19	282.3	132	129
20	295.1	135	131
21	300.1	136	136

Table 4.2 The dataset: a middle-size software project (Peng and Zhai 2017)

Weeks	Cumulative number of detected faults (m_d)	Cumulative number of corrected faults (m_r)
1	12	3
2	23	3
3	43	12
4	64	32
5	84	53
6	97	78
7	109	89
8	111	98
9	112	107
10	114	109
11	116	113
12	123	120
13	126	125
14	128	127
15	132	127
16	141	135
17	144	143

4.4.2   Performance Analysis

To illustrate our models, we consider the following three-paired FDP and FCP models: (1) model with constant debugging lag (abbreviated as M1), (2) model with $\delta(t) = \frac{\ln(1+\gamma t)}{b}$ (abbreviated as M2), and (3) model with exponentially distributed debugging lag (abbreviated M3).

We note that the models proposed in Xie et al. (2007) are special cases of the proposed FDCP and FCP models without considering the dependent faults. For comparison purpose, we also fit the data by the three simplified models of M1–M3 with $p = 1$, which are abbreviated as M1′, M2′, and M3′, respectively.

The six models are fitted to the two datasets by the least squares method. The estimated model parameters for dataset 1 are given in Table 4.3.

As can be noticed from Table 4.1, the estimated parameter a (the total number of faults) in the three proposed models M1–M3 is close to each other. They are all close to 188, which is the number of detected faults after 3-year testing, as reported in Kapur and Younes (1995). On the contrary, the models M1′–M3′, which assume no dependent faults exist, produce quite large a. Therefore, ignoring the dependent faults in the model would result in incorrect total number of faults.

According to the MSE values and the point-wise squared error in Fig. 4.2, it shows that the paired FDP and FCP model with exponentially distributed debugging lag fits the dataset best. On the other hand, the model M1, which assumes constant debugging lag, also provides a compatible fit. The model assuming time-dependent

Model	a	b	Remark	MSE
M1	199.267	0.00717	$\delta = 24.777$	9.0114
			p = 0.3820	
M1$'$	507.474	0.00110	$\delta = 25.705$	10.8924
M2	182.232	0.00955	$\gamma = 0.1599$	15.5697
			p = 0.1374	
M2$'$	1737.637	0.000288	$\gamma = 0.0930$	26.1383
M3	185.148	0.008456	$c = 0.03833$	7.8881
			p = 0.3265	
M3$'$	477.746	0.001177	$c = 0.03786$	10.0985

Table 4.3 The estimated model parameters for dataset 1 (Peng and Zhai 2017)

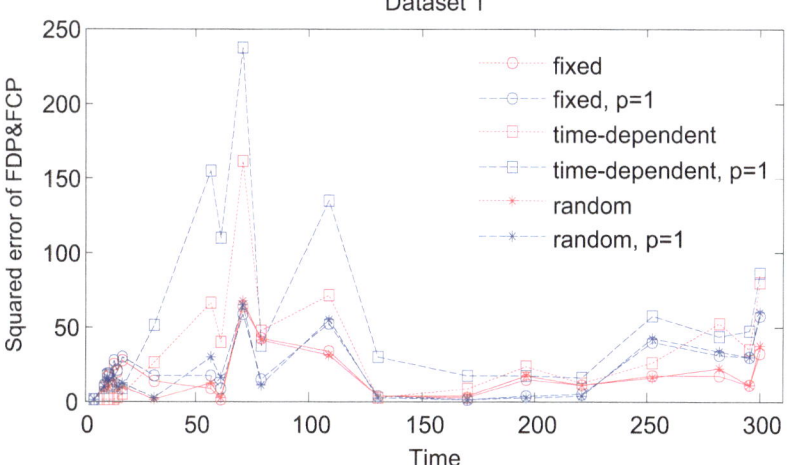

Fig. 4.2 Point-wise squared errors of the six models for dataset 1 (Peng and Zhai 2017)

debugging lags provides the least favorable fit. In fact, according to the estimated model M3, we can infer that the expected length of the debugging lag is $\frac{1}{c} = 26.08$. This is close to the estimated debugging lag in M1. Thus, we can infer that there are significant debugging lags in the software testing process, and it takes about 25 h for a detected fault to be corrected.

The estimation results by the six models for dataset 2 are presented in Table 4.4. Analogous to the dataset 1, the proposed models considering both leading faults and dependent faults are superior to those only considering leading faults. In the fitting procedure, we restrict the total number of faults a to be no smaller than the faults in the data. Therefore, we see that the estimated a are all equal to 144, which is the number of the total faults in dataset 2. Among the three models M1–M3, the constant debugging lag model provides the best fit. This can also be noted from the point-wise squared error in Fig. 4.3. This indicates the debugging lag is almost constant in the software testing process.

Table 4.4 The estimated
model parameters for dataset
2 (Peng and Zhai 2017)

Model	a	b	Remark	MSE
M1	144	0.3058	$\delta = 1.512$	39.5732
			p = 0.474	
M1$'$	153.009	0.1487	$\delta = 1.939$	41.0015
M2	144	0.3938	$\gamma = 0.3112$	49.9352
			p = 0.0448	
M2$'$	168.363	0.1193	$\gamma = 0.0930$	104.8889
M3	144	0.3354	$c = 0.7281$	47.0471
			p = 0.3551	
M3$'$	156.345	0.1404	$c = 0.5811$	55.192

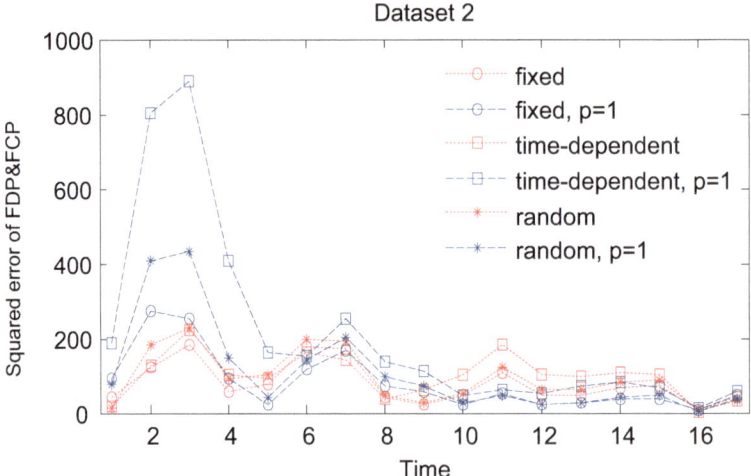

Fig. 4.3 Point-wise squared errors of the six models for dataset 2 (Peng and Zhai 2017)

4.5 Software Release Policy

Developing a SRGM is not the ultimate goal for analysts. Rather, from the
constructed models, useful information can be extracted to guide decision-making.
One critical decision for software projects is to determine the optimal release time
(Huang and Lyu 2005; Inoue and Yamada 2007; Kim et al. 2009; Li et al. 2010;
Pietrantuono et al. 2010). Many studies have dealt with this problem; see Jain and
Priya (2005) and Boland and Chuí (2007) for an overview. As cost and reliability
requirements are of great concern, they are often used as objectives to determine the
time to stop testing and release software (Shinohara et al. 1998; Zhang and Pham
2000, 2002; Jha et al. 2009).

4.5.1   *Software Release Policy Based on Reliability Criterion*

Software is usually released when a reliability target is achieved. It is reasonable to stop testing when a pre-specified proportion of faults are removed. We use T to denote the length of testing and consider the ratio of cumulative removed faults to the initial faults in the software system as the reliability criterion (Huang and Lin 2006):

$$R_1(T) = \frac{m_r(T)}{a}. \tag{4.24}$$

If the reliability target is R_1 and T_1 is the time when the reliability target is reached, we have

$$T_1 = m_r^{-1}(a \cdot R_1). \tag{4.25}$$

Another criterion is the software reliability, which is defined as the probability that no failure occurs during time interval $(T,T+\Delta T]$ given that the software is released at time T. Considering that software normally does not change in operational phase, the reliability function is

$$R(\Delta T|T) = \exp[-\lambda_d(T)\Delta T], \tag{4.26}$$

where $\lambda_d(T)$ is the instantaneous failure intensity at time T. If R_2 is the reliability target for $R(\Delta T|T)$ and T_2 is the time when the reliability of the software reaches R_2, T_2 can be solved by minimizing T subject to $R(\Delta T|T) \geq R_2$.

4.5.2   *Software Release Policy Based on Cost Criterion*

For a basic FDP model with mean value function $m(t)$, the optimal release time T_c can be determined by minimizing the following frequently used cost model (Xie 1991)

$$C(T) = c_1 m(T) + c_2(m(\infty) - m(T)) + c_3 T, \tag{4.27}$$

where c_1 is the expected cost of removing a fault during testing, c_2 is the expected cost of removing a fault in the field, and c_3 is the expected cost per unit time of testing. In practice, the cost of removing a fault in field is always greater than the cost of removing a fault during testing; thus, we assume $c_2 > c_1$.

Analogously, the cost model under the proposed framework with the incorporation of FCP $m_r(t)$ can be expressed as

$$C(T) = c_1 m_r(T) + c_2(m_d(\infty) - m_r(T)) + c_3 T, \tag{4.28}$$

where $m_r(T)$ is the total number of corrected faults at the time of release T and $m_d(\infty) - m_r(T)$ is the number of uncorrected faults that include both the number of undetected faults $m_d(\infty) - m_d(T)$ and the number of detected but uncorrected faults $m_d(T) - m_r(T)$. By minimizing the cost model with respect to T, a more practical optimal release time T_c can be obtained.

Theorem 1 Under the proposed models, the time T_C which minimizes $C(T)$ exists. Specifically, there exist 2k nonnegative numbers $z_1 \leq z_2 \leq \ldots \leq z_{2k}$ which satisfy that $C(T)$ increases on $[0, z_1]$, decreases on $(z_1, z_2), \ldots$, decreases on (z_{2k-1}, z_{2k}), and increases on $[z_{2k}, \infty)$. T_c can be determined as $T_c = \arg\min_{T \in \{0, z_2, \ldots, z_{2k}\}} C(T)$.

Proof We just need to prove that there exists a T_s which satisfies such that $C'(T)$ is positive when T is greater than T_s. Since $C(T) = c_1 m_r(T) + c_2(m_d(\infty) - m_r(T)) + c_3 T$, we have

$$C'(T) = c_3 - (c_2 - c_1)\lambda_r(T). \tag{4.29}$$

Clearly, $C'(0) = c_3 > 0$, , indicating that $C(T)$ is increasing when T is close to zero. We shall prove that $\lambda_r(T)$ approaches 0 (or $C'(T)$ approaches c_3) when T approaches infinity. If so, $C(T)$ is increasing when T is close to 0 or approaches $+\infty$. Consequently, if $C(T)$ has any stationary point, it must have even number of stationary points $z_1 \leq z_2 \leq \ldots \leq z_{2k}$ that satisfy that $C(T)$ increases on $[0, z_1]$, decreases on $(z_1, z_2), \ldots$, decreases on (z_{2k-1}, z_{2k}), and increases on $[z_{2k}, \infty)$. In the following, we show that $\lambda_r(T)$ approaches 0 when T approaches $+\infty$.

If the paired model under constant debugging lag assumption is used, from (4.16) we have

$$\lambda_r(T) = pab e^{b\delta - bT} + ab(1-p)(bT - 2b\delta)e^{2b\delta - bT}, T \geq 2\delta. \tag{4.30}$$

When T approaches $+\infty$, $\lambda_r(T)$ approaches 0.

If the paired model under time-dependent debugging lag assumption is used, from $m_r(t) = m_d(t - b^{-1} \ln(1 + ct))$, we have $\lambda_r(T) = \lambda_d(T - b^{-1} \ln(1 + cT))(1 - b^{-1}c/(1 + cT))$. Based on (4.18), we have

$$\lambda_d(T) = abe^{-bT} - a(1-p)\left(b + bcT - b^2 T - \frac{b^2 cT^2}{2}\right)e^{-bT} \tag{4.31}$$

It can be seen that $\lambda_d(T)$ approaches 0 when T approaches $+\infty$. In addition, $T - b^{-1} \ln(1 + cT)$ approaches $+\infty$ when T approaches $+\infty$. Thus, $\lambda_r(T)$ approaches 0 when T approaches $+\infty$.

If the paired model under exponentially distributed random debugging lag assumption is used, we have

$$\lambda_r(T) = c(m_d(T) - m_r(T)). \tag{4.32}$$

When T approaches $+\infty$ both $m_d(T)$ and $m_r(T)$ approach a, thus, $\lambda_r(T)$ approaches 0 when T approaches $+\infty$.

4.5.3   Software Release Policy Based on Mixed Criterion

When both reliability requirements and the total cost are considered, our goal is to determine the optimal release time T^* which minimizes the total cost without violating the reliability requirements. Therefore, the problem can be formulated as

$$\text{Minimize } C(T) = c_1 m_r(T) + c_2(m_d(\infty) - m_r(T)) + c_3 T$$

$$\text{Subject to } R_1(T) = \frac{m_r(T)}{a} \geq R_1 \ (\text{or } R(\Delta T|T) = \exp[-\lambda_d(T)\Delta T] \geq R_2)$$

1. Considering Reliability Target R_1 and Cost Criterion

   We can divide the time axis $[0, \infty)$ into two types of intervals such that $C(T)$ increases on type 1 intervals and $C(T)$ decreases on type 2 intervals. The candidates for T^* comprise of the minimum T on each type 1 interval that satisfies $R_1(T) \geq R_1$ (or $T \geq T_1$). T^* is the candidate which leads to the lowest cost.

2. Considering Reliability Target R_2 and Cost Criterion

   In this case, we divide the time axis $[0, \infty)$ into four types of intervals such that both $R(\Delta T|T)$ and $C(T)$ increase on type 1 intervals, both $R(\Delta T|T)$ and $C(T)$ decrease on type 2 intervals, and $R(\Delta T|T)$ increases, while $C(T)$ decreases on type 3 intervals, and $R(\Delta T|T)$ decreases, while $C(T)$ increases on type 4 intervals. The candidates for T^* comprise of the minimum T in each type 1 interval that satisfies $R(\Delta T|T) \geq R_2$, the maximum T in each type 2 interval that satisfies $R(\Delta T|T) \geq R_2$, the end points of type 3 intervals which satisfy $R(\Delta T|T) \geq R_2$, and the initial points of type 4 intervals which satisfy $R(\Delta T|T) \geq R_2$. T^* is the candidate which leads to the lowest cost.

4.5.4   Numerical Examples

For illustration, we consider the paired FDP and FCP model with constant debugging lag that fits the dataset 1 in Sect. 4.1. The model parameters are

$a = 186.25$, $b = 0.008$, $\delta = 25.082$, and $p = 0.357$. In addition, we assume $c_1 = \$300$, $c_2 = \$2000$, $c_3 = \$10$, $\Delta T = 12$, $R_1 = 0.95$, and $R_2 = 0.95$.

1. Considering cost criterion and reliability target R_1

   Solving $R_1(T) = m_r(T)/a = 0.95$ gives $T_1 = 588.28$. From (4.29) we have

$$
\begin{aligned}
C(T) &= c_1 m_r(T) + c_2(m_d(\infty) - m_r(T)) + c_3 T \\
&= 372{,}500 - 1700\, m_r(T) + 10T.
\end{aligned}
\tag{4.33}
$$

On the other hand, the correction process model with given parameters is

$$
m_r(T) = \begin{cases}
0, T < 24.78 \\
76.12 - 90.92e^{-0.00717T}, 24.78 \le T < 49.56 \\
199.27 - (204.18 + 1.26T)e^{-0.00717T}, T \ge 49.56
\end{cases}.
$$

By substituting $m_r(T)$ into (4.33), it can be derived that $C(T)$ increases on $[0, 24.78]$, decreases on $(24.78, 1030.45)$, and increases on $[1030.45, \infty)$. As can be verified, $R_1(0) < 0.95$, $R_1(1030.45) > 0.95$. According to the analysis in the preceding section, the optimal release time is $T_1^* = 1030.45$. Correspondingly, the optimal software testing cost is $C(T_1^*) = 45{,}624.87$.

2. Considering cost criterion and reliability target R_2

   When $R_2(\Delta T | T)$ is used as the reliability constraint, we can derive the following detection rate according to the specified model parameters

$$
\lambda_d(T) = \begin{cases}
0.5458e^{-0.00717T}, T < 24.78 \\
(0.3584 + 0.0076T)e^{-0.00717T}, T \ge 24.78
\end{cases}.
$$

It can be verified that $\lambda_d(T)$ decreases on $[0, 24.78)$, increases on $[24.78, 92.07)$, and decreases on $[92.07, \infty)$. Accordingly, $R_2(\Delta T | T)$ increases on $[0, 24.78)$, decreases on $[24.78, 92.07)$, and increases on $[92.07, \infty)$. Referring to the analysis in Sect. 4.5.3, the axis $[0, \infty)$ can be divided into a type 1 interval $[0, 24.78)$, a type 2 interval $[24.78, 92.07)$, a type 3 interval $[92.07, 1030.45$), and a type 1 interval $[1030.45, +\infty)$. Because $R_2(\Delta T | T) < 0.95$ for all $T \in [0, 92.07)$, there is no permissible T in this interval. Therefore, the candidates for the optimal T^* are the right end point 1030.45 of the type 3 interval and the minimum T in $[1030.45, +\infty)$ that satisfies $R_2(\Delta T | T) \ge 0.95$. Because $R_2(\Delta T | 1030.45) = 0.93 < 0.95$, the optimal T_2^* is found as $\arg\min_T\{T \ge 1030.45, R(\Delta T | T) \ge 0.95\} = 1056.81$. The optimal software testing cost is $C(T_2^*) = 45645.38$, which is slightly larger than that in the last case. An illustration of the optimal release policies under two scenarios is given in Fig. 4.4.

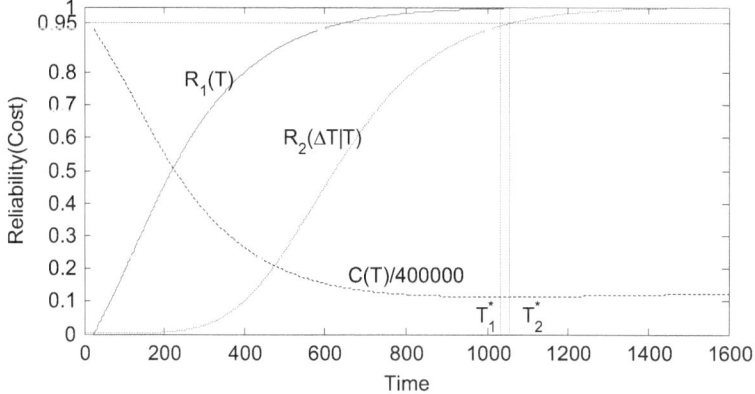

Fig. 4.4 Normalized total cost function and the software reliability functions (Peng and Zhai 2017)

4.6 Summary

In this chapter, modeling of fault detection process and correction process with consideration of fault dependency was investigated. We categorized faults during software testing into leading faults and dependent faults according to their detectability. Leading faults can be detected and corrected directly, whereas dependent faults can only be detected until the corresponding leading faults are corrected. For both types of faults, the FCP is modeled as debugging lags after FDP, and the dependency of the FDP of dependent faults on the FCP of leading faults was considered. Several paired models were derived based on different assumptions on the debugging lag. The obtained paired model under constant debugging lag assumption was fitted to two real datasets for illustration. Under this framework, optimal software release policy was comprehensively studied.

The proposed modeling framework can be further extended to incorporate other information or to adapt to other testing environments. For instance, Bayesian technique can be used to incorporate prior information and update model parameters when more information is available. In addition, the imperfect fault correction or the fault introduction phenomenon can be incorporated, as it is common for debuggers to make mistakes with fault correction.

References

Boland, P. J., & Chuí, N. N. (2007). Optimal times for software release when repair is imperfect. *Statistics & Probability Letters, 77*, 1176–1184.

Chang, Y. C., & Liu, C. T. (2009). A generalized JM model with applications to imperfect debugging in software reliability. *Applied Mathematical Modelling, 33*, 3578–3588.

Goel, A. L., & Okumoto, K. (1979). Time-dependent error-detection rate model for software reliability and other performance measures. *IEEE Transactions on Reliability, 28*(3), 206–211.

Gokhale, S. S., Lyu, M. R., & Trivedi, K. S. (2004). Analysis of software fault removal policies using a non-homogeneous continuous time Markov chain. *Software Quality Journal, 12*(3), 211–230.

Goseva-Popstojanova, K., & Trivedi, K. S. (2000). Failure correlation in software reliability models. *IEEE Transactions on Software Engineering, 49*(37), 37–47.

Gutjahr, W. J. (2001). A reliability model for nonhomogeneous redundant software versions with correlated failures. *Computer Systems Science and Engineering, 16*(6), 361–371.

Hu, Q. P., Xie, M., Ng, S. H., & Levitin, G. (2007). Robust recurrent neural network modeling for software fault detection and correction prediction. *Reliability Engineering & System Safety, 92*(3), 332–340.

Huang, C. Y., & Huang, W. C. (2008). Software reliability analysis and measurement using finite and infinite server queueing models. *IEEE Transactions on Reliability, 57*(1), 192–203.

Huang, C. Y., & Lin, C. T. (2006). Software reliability analysis by considering fault dependency and debugging time lag. *IEEE Transactions on Reliability, 55*(3), 436–450.

Huang, C. Y., & Lyu, M. R. (2005). Optimal release time for software systems considering cost, testing effort, and testing efficiency. *IEEE Transactions on Reliability, 54*(4), 583–591.

Huang, C. Y., Kuo, S. Y., & Lyu, M. R. (2007). An assessment of testing-effort dependent software reliability growth models. *IEEE Transactions on Reliability, 56*(2), 198–211.

Inoue, S., & Yamada, S. (2007). Generalized discrete software reliability modeling with effect of program size. *IEEE Transactions on Systems, Man, and Cybernetics: Part A-Systems and Humans, 37*(2), 170–179.

Jain, M., & Priya, K. (2005). Software reliability issues under operational and testing constraints. *Asia-Pacific Journal of Operational Research, 22*(1), 33–49.

Jha, P. C., Gupta, D., Yang, B., & Kapur, P. K. (2009). Optimal testing resource allocation during module testing considering cost, testing effort and reliability. *Computers and Industrial Engineering, 57*(3), 1122–1130.

Jia, L. X., Yang, B., Guo, S. C., & Park, D. H. (2010). Software reliability modeling considering fault correction process. *IEICE Transactions on Information and Systems, E93D*(1), 185–188.

Kapur, P. K., & Younes, S. (1995). Software reliability growth model with error dependency. *Microelectronics Reliability, 35*(2), 273–278.

Kapur, P. K., Goswami, D. N., Bardhan, A., & Singh, O. (2008). Flexible software reliability growth model with testing effort dependent learning process. *Applied Mathematical Modeling, 32*, 1298–1307.

Kim, H. S., Park, D. H., & Yamada, S. (2009). Bayesian optimal release time based on inflection s-shaped software reliability growth model. *IEICE Transactions on Fundamentals of Electronics, Communications and Computer Sciences, E92A*(6), 1485–1493.

Lee, C. H., Kim, Y. T., & Park, D. H. (2004). S-shaped software reliability growth models derived from stochastic differential equations. *IIE Transactions, 36*(12), 1193–1199.

Li, X., Xie, M., & Ng, S. H. (2010). Sensitivity analysis of release time of software reliability models incorporating testing effort with multiple change-points. *Applied Mathematical Modelling, 34*(11), 3560–3570.

Lin, C. T., & Huang, C. Y. (2008). Enhancing and measuring the predictive capabilities of testing-effort dependent software reliability models. *The Journal of Systems and Software, 81*, 1025–1038.

Lin, C. T., & Huang, C. Y. (2009). Staffing level and cost analyses for software debugging activities through rate-based simulation approaches. *IEEE Transactions on Reliability, 58*(4), 711–724.

Lin, C. T., & Li, Y. F. (2014). Rate-based queueing simulation model of open source software debugging activities. *IEEE Transactions on Software Engineering, 40*(11), 1075–1099.

Lo, J. H., & Huang, C. Y. (2006). An integration of fault detection and correction processes in software reliability analysis. *The Journal of Systems and Software, 79*, 1312–1323.

Lyu, M. R. (1996). *Handbook of software reliability engineering.* New York: McGraw Hill.

Musa, J. D., Iannino, A., & Okumono, K. (1987). *Software reliability, measurement, prediction and application.* New York: McGraw-Hill.

Okamura, H., & Dohi, T. (2011). Unification of software reliability models using markovian arrival processes. In *17th Pacific Rim International Symposium on Dependable Computing (PRDC)* (pp. 20–27). IEEE.

Okamura, H., Dohi, T., & Osaki, S. (2013). Software reliability growth models with normal failure time distributions. *Reliability Engineering & System Safety, 116*, 135–141.

Peng, R., & Zhai, Q. (2017). Modeling of software fault detection and correction processes with fault dependency. *Eksploatacja i Niezawodnosc– Maintenance and Reliability, 19*(3), 467–475.

Peng, R., Li, Y. F., Zhang, W. J., & Hu, Q. P. (2014). Testing effort dependent software reliability model for imperfect debugging process considering both detection and correction. *Reliability Engineering and System Safety, 126*, 37–43.

Peng, R., Li, Y. F., Zhang, J. G., & Li, X. (2015). A risk-reduction approach for optimal software release time determination with the delay incurred cost. *International Journal of Systems Science, 46*(9), 1628–1637.

Pietrantuono, R., Russo, S., & Trivedi, K. S. (2010). Software reliability and testing time allocation: An architecture-based approach. *IEEE Transactions on Software Engineering, 36*(3), 323–337.

Schneidewind, N. F. (1975). *Analysis of error processes in computer software, Proceedings of the International Conference on Reliable Software* (pp. 337–346). Los Alamitos: IEEE Computer Society Press.

Shatnawi, O. (2009). Discrete time modelling in software reliability engineering-a unified approach. *Computer Systems Science and Engineering, 24*(6), 391–397.

Shatnawi, O. (2014). Measuring commercial software operational reliability: An interdisciplinary modelling approach. *Eksploatacja i Niezawodnosc-Maintenance and Reliability, 16*(4), 585–594.

Shibata, K., Rinsaka, K., Dohi, T., & Okamura, H. (2007). *Quantifying software maintainability based on a fault-detection/correction model*. 13th Pacific Rim International Symposium on Dependable Computing (PRDC 2007), IEEE, pp. 35–42.

Shinohara, Y., Nishio, Y., Dohi, T., & Osaki, S. (1998). An optimal software release problem under cost rate criterion: Artificial neural network approach. *Journal of Quality in Maintenance Engineering, 4*(4), 236–247.

Tamura, Y., & Yamada, S. (2006). A flexible stochastic differential equation model in distributed development environment. *European Journal of Operational Research, 168*, 143–152.

Wang, L., Hu, Q., & Liu, J. (2016). Software reliability growth modeling and analysis with dual fault detection and correction processes. *IIE Transactions, 48*(4), 359–370.

Wu, Y. P., Hu, Q. P., Xie, M., & Ng, S. H. (2008). Modeling and analysis of software fault detection and correction process by considering time dependency. *IEEE Transactions on Reliability, 56*(4), 629–642.

Xie, M. (1991). *Software reliability modeling*. Singapore: World Scientific Publishing Company.

Xie, M., Hu, Q. P., Wu, Y. P., & Ng, S. H. (2007). A study of the modeling and analysis of software fault-detection and fault-correction processes. *Quality and Reliability Engineering International, 23*, 459–470.

Yamada, S., Ohba, M., & Osaki, S. (1984). S-shaped reliability growth models and their applications. *IEEE Transactions on Reliability, 33*, 289–292.

Yang, B., Guo, S., Ning, N., & Huang, H. Z. (2008). Parameter estimation for software reliability models considering failure correlation. In *Annual Reliability and Maintainability Symposium (RAMS 2008)* (pp. 405–410). IEEE.

Zhang, X. M., & Pham, H. (2000). Comparisons of nonhomogenuous Poisson process software reliability models and its applications. *International Journal of Systems Science, 31*(9), 1115–1123.

Zhang, X. M., & Pham, H. (2002). Predicting operational software availability and its applications to telecommunication systems. *International Journal of Systems Science, 33*(11), 923–930.

Zhang, X. M., Teng, X. L., & Pham, H. (2003). Considering fault removal efficiency in software reliability assessment. *IEEE Transactions on Systems, Man and Cybernetics: Part A-Systems and Humans, 33*(1), 114–120.

Zhao, J., Liu, H. W., Cui, G., & Yang, X. Z. (2006). Software reliability growth model with change-point and environmental function. *The Journal of Systems and Software, 79*(11), 1578–1587.

Chapter 5
General Order Statistics-Based Model

In this chapter, a framework of software reliability models containing both information from software fault detection process and correction process is studied. Different from previous chapters, the proposed approach is based on Markov model other than a nonhomogeneous Poisson process model. Also, parameter estimation is carried out with weighted least-square estimation method which emphasizes the influence of later data on the prediction. Datasets from practical software development projects are applied with the proposed framework, which shows satisfactory performance with the results.

5.1 Introduction

In the past few decades, the software reliability growth models (SRGMs) have attracted widespread research attention and are applied in the software development process (Lyu 2008; Xie 1993). Focusing on the software development life cycles (Dyer 1991) and via black box testing method, most SRGMs utilize the fault-detected data during testing process to describe the stochastic behavior of the software fault detection process (FDP) with respect to time, in which the most popular parametric models are the nonhomogeneous Poisson process (NHPP) models (Goel 1985; Lyu 1996; Okamura et al. 2013; Yamada and Fujiwara 2001) and Markov models (Jelinski and Moranda 1972; Xie 1993; Aktekin and Caglar 2013; Huang et al. 2014). With more and more SRGMs being proposed and studied, a unification research is directed to establish a unified modeling framework comprised of typical reliability growth patterns. Some unification schemes are proposed, such as a generalized order statistic (GOS) model (Miller 1986), a semi-parametric model which is put forward by Ross (Ross 1985), Kapur's generalized imperfect debugging NHPP models (Kapur et al. 2011), and a unified modeling framework of Markov-type software reliability models (Okamura and

© The Author(s), under exclusive license to Springer Nature Singapore Pte Ltd. 2018   53
R. Peng et al., *Software Fault Detection and Correction: Modeling and Applications*,
SpringerBriefs in Computer Science, https://doi.org/10.1007/978-981-13-1162-8_5

Dohi 2011). Similar to the GOS models, a novel model is developed which ideally assumes each fault in the software independently causing errors to occur with exponential order statistic process whose error intensity is unknown (Koh 2008).

However, most of these modeling frameworks are under the idealized assumptions that all the faults are removed instantaneously and perfectly, and the expected number of removed faults is the same as the expected number of detected faults. In fact, it always needs time for fault removal, and the expected number of faults removed at any given time is less than the expected number of detected faults (Gokhale et al. 2004). Since then, some researchers also took into account of the fault correction process (FCP) and used the fault-corrected data to represent the correction time delay. Modeling both FDP and FCP takes more information from software testing records and improves the estimation and prediction results. For instance, Schneidewind proposed an approach for modeling the fault correction process by using a constant delayed fault detection process (Schneidewind 2009). He assumed that the rate of fault correction was proportional to the rate of fault detection. However, since the FCP depends heavily on the FDP, in some applications, the model will underestimate the remaining faults in the code. Later, Xie et al. extended the Schneidewind model to a continuous version by substituting a time-dependent delay function for the constant delay (Xie et al. 2007; Wu et al. 2008). Based on the proposed FDP and FCP modeling framework and NHPP assumption, both the least-square estimation and maximum likelihood estimation approaches have been proposed in the research. Some researchers incorporated the FCP with a queueing theory and assumed an infinite number of servers (Shibata et al. 2007; Huang and Hung 2010). Later, a neural networks configuration approach with an extra factor characterizing the dispersion of prediction repetitions is applied to model these two processes together (Hu et al. 2007). Recently, incorporating testing-effort function and imperfect debugging, Peng proposed a framework to analyze those processes (Peng et al. 2014). On the other hand, extensions on the Markov models have also been conducted by Gokhale in a systematic way (Gokhale et al. 2004). The Markov model-based FDP and FCP modeling framework incorporates both the non-zero fault correction time and fault introduction factors. Analysis for this framework is conducted in a simulated way, which is applied for the decision-making of fault correction policies.

In this chapter, a novel FDP and FCP modeling framework based on the GOS models is proposed, with both the modeling and statistical analysis methods. Based on the traditional GOS models for FDP, the FCP is regarded as a delayed one. Apart from FDP and FCP modeling, the parameter estimation is also an important issue in research. It is well accepted that the MLE has many desirable properties, such as asymptotic-normality, admissibility, and consistency (Sahinoglu and Can 1997), and some complex likelihood functions for combined FDP and FCP are developed (Wu et al. 2008; Shibata et al. 2007). However, ML has some inherent defects (Ishii et al. 2012) and can only be applied to detection and correction data with an equal time interval. In the Jelinski-Moranda model which is the earliest GOS model, the MLE does not necessarily exist, and it is not necessarily unique (Jukic 2012). The ordinary least-square (OLS) estimation method is a more common method to estimate parameters due to its simplicity, while the OLS estimate lacks a sound theoretical explanation and ignores the different variance of each data point (Draper and Smith 1998). To improve OLS estimate and try to balance the heteroscedasticity, which is introduced by the testing mechanism, the weighted least-square (WLS) estimation is applied here.

The rest of this chapter is organized as follows. Section 5.4.2 outlines the unification framework for FDP and FCP modeling. In Sect. 5.4.3 the parameter estimation with WLS estimate is proposed and developed. Section 5.4.4 illustrates the application of proposed FDP and FCP models with two practical datasets. And then, reliability analysis and optimal release time issues are addressed.

5.2 Modeling Framework

Assumptions

1. Based on the debugging theory mentioned in Ohishi et al. (2009), assume that the software faults occur at independent distributed random times, as a fact of the unknown location of fault.
2. The initial number of the faults N is positive and finite. This is a common assumption that has been used in most of software reliability growth model (Lyu 1996; Huang et al. 2003), which can be easily relaxed by introducing more model parameters.
3. When a fault remains in the code, its state remains in 0. When the i-th fault is detected, its state changes to 1 with detection rate, and when it is corrected, its state changes to 2 with correction rate.

The fault state transition process is as shown in Fig. 5.1.

The fault detection rate is related to the debuggers' detection capacity, the test case testability, and test coverage. The fault correction rate is related to the debuggers' correction capacity and the complexity of each fault. In this assumption, this is reasonable to assume each detected fault will enter the correction process immediately, because during the test of large-scale software, the fault detection and correction are conducted in parallel.

Those general assumptions allow us to incorporate the fault correction data into modeling and easily develop more specific models suitable in different situations. More parameters can be added for more flexible use.

5.2.1 Modeling Fault Detection Process

Let λ_i denotes the error detection intensity of the i-th fault, and $I_{di}(t)$ is the indicator function of the i-th fault's state, which

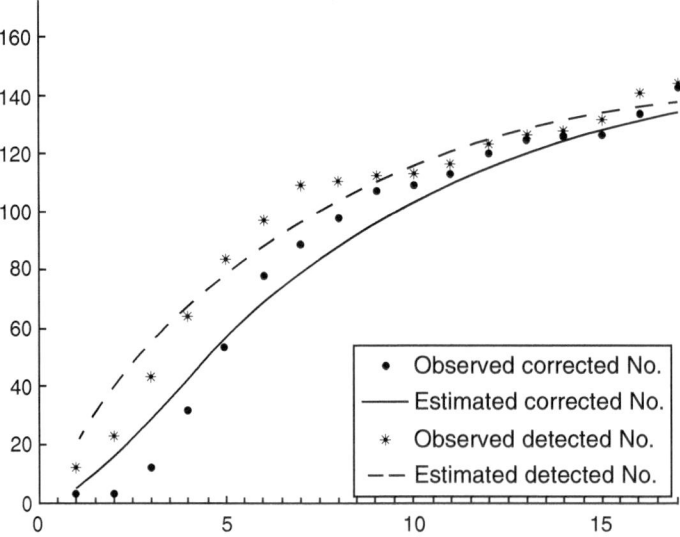

Fig. 5.1 Fault state transition process ©[2016] John Wiley and Sons. Reprinted, with permission, from Liu et al. (2016)

$$I_{di}(t) = \begin{cases} 1, & \text{Fault has not been detected} \\ 0, & \text{Otherwise} \end{cases}.$$

Then the *i-th* fault-detected probability density function (PDF) obeys, for example, exponential distribution, and can be described as follows:

$$f_{di}(t) = \lambda_i e^{-\lambda_i t}. \tag{5.1}$$

Since the error rate of every fault is λ_i, the expected error rate of the software $\Lambda_d(t)$ is

$$E[\Lambda_d(t)] = \sum_{i=1}^{N} \lambda_i E[I_{di}(t)] = \sum_{i=1}^{N} \lambda_i \int_t^{\infty} f_{di}(\tau)d\tau = \sum_{i=1}^{N} \lambda_i e^{-\lambda_i t}. \tag{5.2}$$

Then, the mean value function (MVF) of fault detection number is formulated as follows:

$$m_d(t) = E(N_d(t)) = \sum_{i=1}^{N} [1 - \exp(-\lambda_i t)]. \tag{5.3}$$

The variance of the detected fault number by time t presented in (5.3) can be derived as

$$Var(N_d(t)) = \sum_{i=1}^{N}[Var(1 - I_{di}(t))]$$

$$= \sum_{i=1}^{N}[Var(I_{di}(t))] = \sum_{i=1}^{N}\left[E(I_{di}(t)) - E^2(I_{di}(t))\right] \quad (5.4)$$

For mathematical purposes, the potential number of initial faults N is viewed as real value even though it is really integer valued. And if it assumes that $\lambda_i = \lambda_0$, $i = 1$, 2, ..., N (Miller 1986), the MVF can be simplified to the famous Jelinski-Moranda (J-M) model (Jelinski and Moranda 1972):

$$m_d(t) = E(N_d(t)) = N(1 - \exp(-\lambda_0 t)). \quad (5.5)$$

For the J-M model, the variance of the detected fault number by time t is as follows, which has been proposed before by Koh (Koh 2008):

$$Var(N_d(t)) = N(\exp(-\lambda_0 \cdot t) - \exp(-2\lambda_0 \cdot t)) \quad (5.6)$$

5.2.2 Modeling Fault Correction Process

To model FCP, assume that μ_i is the error correction intensity of the i-th fault that has been detected and $I_{ci}(t) = 0$ denotes the i-th fault has been corrected prior to t and $I_{ci}(t) = 1$; if not, then

$$E(I_{ci}(t)) = P(t_d + \Delta t_c > t), \quad (5.7)$$

where t_d and Δt_c refer to the duration of fault detection and correction.

According to above assumptions and using exponential distribution which is applied in fitting random fault correction time in many practical software testing projects (Musa et al. 1987), the PDF of successfully correcting *i-th* fault can be described as follows:

$$f_{\Delta ci}(\Delta t) = \mu_i \exp(-\mu_i \cdot \Delta t). \quad (5.8)$$

Since every fault will enter FCP immediately after being detected, then the error correction process can be regarded as an overlay of two GOS processes as follows:

$$E(I_{ci}(t)) = \int_t^\infty \int_0^\tau (f_{di}{}^* f_{\Delta ci})(\tau)d\tau = \int_t^\infty \int_0^\tau f_{di}(x)f_{\Delta ci}(\tau - x)dxd\tau$$

$$= \begin{cases} \dfrac{\mu_i e^{-\lambda_i t} - \lambda_i e^{-\mu_i t}}{\mu_i - \lambda_i}, & \lambda_i \neq \mu_i \\ (\lambda_i t + 1)e^{-\lambda_i t} & , \lambda_i = \mu_i \end{cases}, \quad \lambda_i \ \mu_i > 0. \tag{5.9}$$

Similar to the derivation of MVF in FDP, the MVF in FCP can be derived as

$$N_c(t) = \sum_{i=1}^N (1 - I_{ci}(t)), \tag{5.10}$$

$$E(N_c(t)) = \sum_{i=1}^N E(1 - I_{ci}(t)) = \sum_{i=1}^N (1 - E(I_{ci}(t))). \tag{5.11}$$

If combining with the J-M model and assuming $\mu_i = \mu_0$, $i = 1, 2, \ldots, N$, the MVF of FCP is formulated as follows:

$$m_c(t) = E[N_c(t)] = \begin{cases} N\left(1 - \dfrac{\mu_0 e^{-\lambda_0 t} - \lambda_0 e^{-\mu_0 t}}{\mu_0 - \lambda_0}\right) & , \lambda_0 \neq \mu_0 \\ N\left(1 - (\lambda_0 t + 1)e^{-\lambda_0 t}\right) & , \lambda_0 = \mu_0 \end{cases}, \quad \lambda_0 \ \mu_0 > 0. \tag{5.12}$$

The above formulation is similar to the FCP model proposed by Wu et al. (2008) which follows NHPP. The most obvious difference between these two models is that the GOS model supposes the number of initially contained faults is finite while the NHPP model assumes it as a Poisson distributed random variable with the mean value of N (Langberg and Singpurwalla 1985). Aside from this, their variances of the expected cumulated number of faults have different tendency, which will be illustrated in following content.

With different assumptions on the distribution of correction time delay and the detection rate, the proposed model framework can have different formats. The different formats with the exponential detection rate are similar with the models proposed by Wu et al. 2008).

The variance of $N_c(t)$ is derived as

$$Var(N_c(t)) = \sum_{i=1}^N \left[E(I_{ci}^2(t)) - E^2(I_{ci}(t)) \right]$$

$$= N - \left[\sum_{i=1}^N E^2(I_{ci}(t)) \right] - m_c(t). \tag{5.13}$$

When describing the FDP and FCP with the model shown in (5.12), the variance of $N_c(t)$ can be simplified as follows:

$$Var(N_c(t)) = N - \frac{(N - m_c(t))^2}{N} - m_c(t). \tag{5.14}$$

According to (5.14), it can be easily proved that the expected variance of $N_c(t)$ increases with testing time t when $m_c(t) < N/2$ and decreases when $m_c(t) \geq N/2$.

Above all, the proposed model is different from the existed model based on NHPP (Xie et al. 2007; Wu et al. 2008; Shibata et al. 2007; Huang and Hung 2010) for the following reasons:

1. Proposed model is based on general order statistics, and the initial fault count embedded in software is fixed, while the NHPP-based model is assumed the initial fault count is distributed as Poisson distribution.
2. The MVF of proposed model has a decreasing variance when t is big enough, and the MVF of NHPP-based model always possesses an increasing variance with t, which is

$$Var_d^{(\text{NHPP})}(t) = m_d(t), \quad Var_c^{(\text{NHPP})}(t) = m_c(t).$$

The proposed modeling framework allows researchers to incorporate the fault correction data into modeling and easily generate various models with different assumptions on fault detection and correction distributions.

5.3   Parameter Estimation with Weighted Least-Square Estimation Method

OLS estimate is simple in calculation and widely use in SRGM, but it lacks a sound theoretical explanation and ignores the different variances of each data point (Draper and Smith 1998), which is introduced by the testing mechanism. To improve OLS and try to balance the variance, the weighted least-square estimation is used here.

Suggested by Koh (2008), construct new non-heteroscedasticity variables independent with t,

$$\begin{cases} \dfrac{n_k - E\left(N_d\left(t_k^d\right)\right)}{\sqrt{Var\left(N_d\left(t_k^d\right)\right)}} = \varepsilon_k^d\left(t_k^d, \theta\right) \\ \dfrac{m_j - E\left(N_c\left(t_j^c\right)\right)}{\sqrt{Var\left(N_c\left(t_j^c\right)\right)}} = \varepsilon_j^c\left(t_j^c, \theta\right) \end{cases} \tag{5.15}$$

in which θ refers to the parameters to be estimated in the model and t_k^d (t_j^c) refers to the k-th (j-th) recording time of FDP (FCP). It is worth noting that in some software testing record, the recording time interval of FDP and FCP can be different from each other, i.e., when $k = j$, not every t_k^d is equal to t_j^c.

It can be proved that the second moment of ε_d and ε_c equal to 1 uses the logarithm of the normalized second central moment of $m_d(t)$ and $m_c(t)$, i.e., $\ln\left(\varepsilon_d^2\right)$ and $\ln\left(\varepsilon_d^2\right)$, to construct the optimal weighted least-square estimation which can be written as

$$
\begin{aligned}
\underset{\theta}{\text{Min}}\, S^2 &= \sum_{k=1}^{N'}\left\{\ln\left[\varepsilon_k^d\left(t_k^d,\theta\right)\right]^2\right\}^2 + \sum_{j=1}^{M'}\left\{\ln\left[\varepsilon_j^c\left(t_j^c,\theta\right)\right]^2\right\}^2 \\
&= \sum_{k=1}^{N'}\left\{2\ln\left|m_d\left(t_k^d,\theta\right) - n_k\right| - \ln\left[Var\left(N_d\left(t_k^d,\theta\right)\right)\right]\right\}^2 \\
&\quad + \sum_{j=1}^{M'}\left\{2\ln\left|m_c\left(t_j^c,\theta\right) - m_j\right| - \ln\left[Var\left(N_c\left(t_j^c,\theta\right)\right)\right]\right\}^2 \\
&= \sum_{k=1}^{N'}\left[2\ln\left(\left|m_d\left(t_k^d\right) - n_k\right|\right) + 2\ln\left(w_d\left(t_k^d\right)\right)\right]^2 \\
&\quad + \sum_{j=1}^{M'}\left[2\ln\left(\left|m_c\left(t_j^c\right) - m_j\right|\right) + 2\ln\left(w_c\left(t_j^c\right)\right)\right]^2
\end{aligned}
\tag{5.16}
$$

in which

$$
w_d(t) = \frac{1}{\sqrt{Var(N_d(t))}},
$$

and

$$
w_c(t) = \frac{1}{\sqrt{Var(N_c(t))}}.
$$

The deposition of variance, $w_d(t)$ and $w_c(t)$, increases with t when t is large enough. This character is also consistent with the important intuition that ($N_d(t)$ or $N_c(t)$) which is observed for large t is more reliable to predict the future ($N_d(t)$ or $N_c(t)$). On the other hand, because of the increasing variance, if the WLS estimation method is applied with NHPP models, more weight will be addressed on the early data points, which will be opponent to the above intuition, and deteriorate the prediction accuracy.

To obtain the minimum of S^2, a set of initial parameters is adopted to the minisearch function in MATLAB, and iterations are executed with different initial parameters decided by the last estimation.

In order to make a comparison of various models' performance, mean squared error (MSE) is used to quantify the difference between actual data and the estimated data. A smaller MSE represents a smaller fitting error and better performance. The MSE is generally defined as

$$
\begin{cases}
MSE_d(I_d) = \dfrac{1}{I_d}\sum_{k=1}^{I_d}\left(\dfrac{m_d\left(t_k^d\right) - n_k}{n_k}\right)^2 \\[4mm]
MSE_c(I_c) = \dfrac{1}{I_c}\sum_{j=I_c+1}^{I_c}\left(\dfrac{m_c\left(t_j^c\right) - m_j}{m_j}\right)^2,
\end{cases}
\tag{5.17}
$$

Here k is the size of the selected data set, b_i is the actual number of (detected or corrected) faults by time t_i, and $m(t_i)$ is the estimated number of (detected or corrected) faults by time t_i.

To study the prediction accuracy of the models, the variable-term prediction is used to quantify this ability (Malaiya et al. 1992). This approach predicts the detected fault number $m_d(t_k)\{k = l_d + 1, l_d + 2, \ldots, N'\}$ and the corrected fault number $m_c(t_j)\{j = l_c + 1, l_c + 2, \ldots, M'\}$, using the partial dataset $\{m_d(t_k)\}$ $\{i = 0, 1, 2 \ldots, l_d\}$ and $m_c(t_j)\{j = 0, 1, 2 \ldots, l_c\}\}$. The measure for these errors is mean of relative errors (MRE), which is

$$
\begin{cases}
MRE_d(l_d) = \dfrac{1}{N' - I_d + 1}\sum_{k=I_d+1}^{N'}\left|\dfrac{m_d\left(t_k^d\right) - n_k}{n_k}\right| \\[4mm]
MRE_c(l_c) = \dfrac{1}{M' - I_c + 1}\sum_{j=I_c+1}^{M'}\left|\dfrac{m_c\left(t_j^c\right) - m_j}{m_j}\right|.
\end{cases}
\tag{5.18}
$$

The MRE of FDP and FCP is calculated separately. The lower MRE indicates the better prediction accuracy.

5.4 Some Application Examples

In this section, two application examples are shown to illustrate the approach for FDP and FCP models and their parameter estimation with WLS method. Some other existed models are compared to illustrate the WLS being an effective method. In the first part, an earlier dataset with equal time interval ($t_k^d = t_j^c$, when $k = j$) is used. In the second part, a new and real application datasets with unequal time interval ($t_k^d \neq t_j^c$, when $k = j$) is applied.

Table 5.1 Dataset published in (Xie et al. 2007) ©[2016] John Wiley and Sons. Reprinted, with permission, from Liu et al. (2016)

Test week	Detected no.	Corrected no.	Test week	Detected no.	Corrected no.
1	12	3	10	114	109
2	23	3	11	116	113
3	43	12	12	123	120
4	64	32	13	126	125
5	84	53	14	128	127
6	97	78	15	132	127
7	109	89	16	141	135
8	111	98	17	144	143

5.4.1   Application to an Existing Dataset

In this part, the dataset published by *Xie* (Xie et al. 2007) is used as an example and shown in Table 5.1. This dataset is grouped testing data with equal time interval in FDP and FCP. Assuming that both of the faults detecting time and correcting duration are exponentially distributed, the proposed modeling framework is described as (5.3) and (5.12). Using WLS estimate, the estimated parameters are $\widehat{N} = 152.8677, \widehat{b} = 0.1429, \widehat{\mu} = 0.5641$. The estimated FDP and FCP are shown in Fig. 5.2. Figure 5.3 illustrates that when $t > 6$ weeks, the weight of FDP and FCP is increasing with t, i.e., more weight is assigned on the latest data errors.

5.4.2   Predictability Comparison

To illustrate the predictability of the proposed model, three different types of models are compared in this part. These models are

Model 1, the GOS-based exponential time delay (ETD) model (proposed model) with WLS estimate
Model 2, the ETD model with OLS estimate
Model 3, the NHPP-based ETD model with WLS estimate

While the assumptions of fault detection process in above three models are quite different from each other, the assumptions of time delay for the fault correction is the same. Thus, the mean value function of each model is in similar format. To justify the assumptions of the proposed framework, those three models are compared.

For simplification, assume each fault-detected distribution obeys the exponential distribution with the same fault detection rate b and the exponential correction time delay with the correction rate μ.

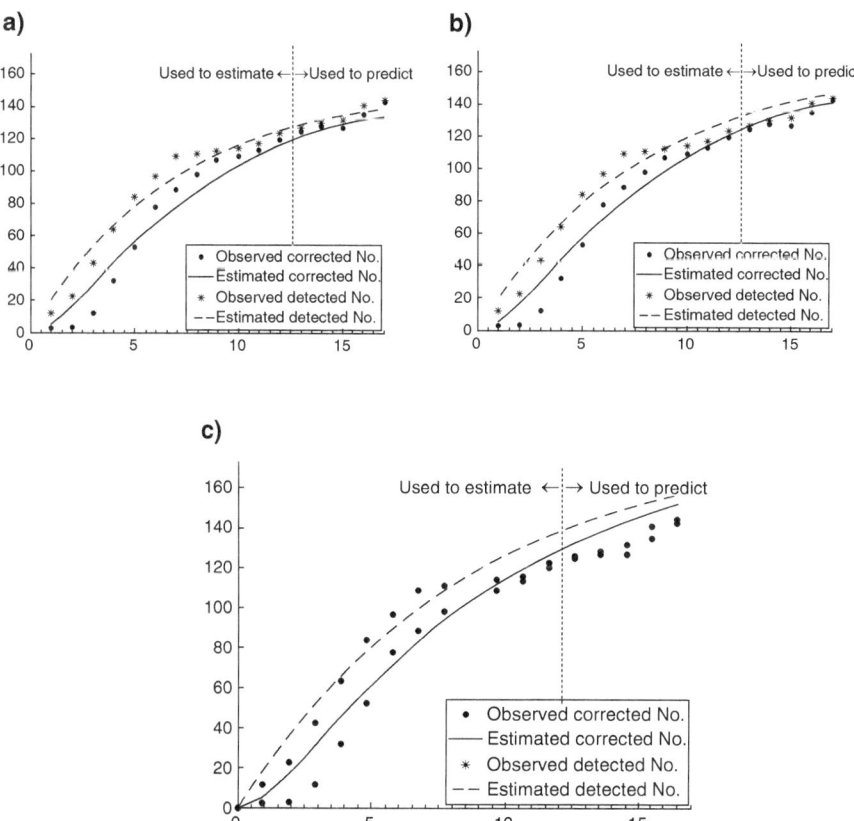

Fig. 5.2 Actual and estimate fault detection and correction processes ©[2016] John Wiley and Sons. Reprinted, with permission, from Liu et al. (2016)

Above models are adopted in the first 12 weeks to predict the remaining 5 weeks. $MRE_d(12)$ and $MRE_c(12)$ are used as measures of prediction ability. The results of estimation and predictability of these models are listed in Table 5.2 and shown in Fig. 5.4. The 95% confidence intervals for model 1 and model 3 are produced by bootstrap algorithm. It shows that the proposed model has the best overall prediction accuracy and achieves much better agreement between actual and estimated FDP. Model 3 did much worse in prediction than other two models, tending to be overprediction on the fault numbers. The results also show that the OLS estimate can lead to overfitting which may influence the predictability.

Comparing the 95% confidence intervals, model 1 has the narrowest interval for N, which is the expected number of initial faults in software. The estimates of N in model 2 and model 3 are located near the upper bound of the estimate in model 1.

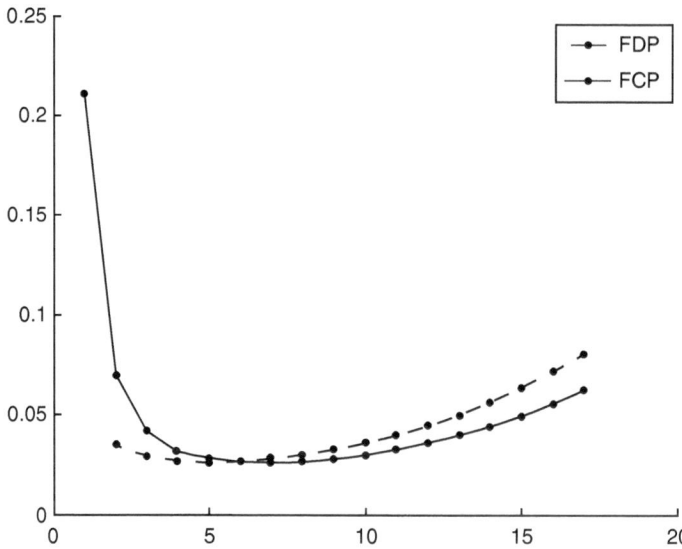

Fig. 5.3 Weight of the proposed model ©[2016] John Wiley and Sons. Reprinted, with permission, from Liu et al. (2016)

5.4.3   Optimal Software Release Time Analysis

One of the most significant applications of software reliability models is the determination of software release time. Some cost models have been proposed in previous works (Chiu et al. 2009). Considering with FCP, the cost model can be written as

$$C(T) = c_{1d} \times m_d(T) + c_{1c} \times m_c(T) + c_2 \times (N - m_c(T)) + c_3 \times T. \qquad (5.19)$$

where c_{1d} and c_{1c} are the expected cost of finding and removing a fault during test phase, since both of the works are carried on separately, c_2 is the cost for the per remaining uncorrected fault, and c_3 is the expected cost per unit time for testing. To confirm the existence of optimal release time of T, c_2 must be larger than c_{1c}, which means removing a remaining uncorrected fault after release will cost more than that during testing. Here, these cost items are assumed as $c_{1d} = 100$, $c_{1c} = 100$, $c_2 = 500$, $c_3 = 100$.

Figure 5.5 shows the plot of the total cost comparing with the proposed model with WLS estimate and ETD model with OLS estimate. The optimal release times for both of the models are 31 with a total cost of 19,090 and 35 with a total cost of 21,052, respectively. Since the WLS estimates have a better prediction performance (shown above), its optimal release time is more accurate than OLS. Also, due to its overfitting phenomenon, the OLS estimation can be a reference to a conservative estimation of software release time which means the cost is more likely to be bigger than real situation.

Table 5.2 Estimates and predictability ©[2016] John Wiley and Sons. Reprinted, with permission, from Liu et al. (2016)

Result		Model		
		Proposed model	Exponential time delay model	NHPP model with WLS
Parameter		$\widehat{N} = 150.9730 \pm 17.4882$ $\widehat{b} = 0.1465 \pm 0.0407$ $\widehat{\mu} = 0.5677 \pm 0.2317$	$\widehat{N} = 165.5349 \pm 40.1480$ $\widehat{b} = 0.1286 \pm 0.0570$ $\widehat{\mu} = 0.5777 \pm 0.2527$	$\widehat{\theta} = 165.4736 \pm 40.4885$ $\widehat{b} = 0.1314 \pm 0.0748$ $\widehat{\mu} = 0.3970 \pm 0.0193$
Prediction	$MRE_d(12)$	0.027	0.053	0.060
	$MRE_c(12)$	0.030	0.025	0.020
	MRE_{all}	0.057	0.078	0.080

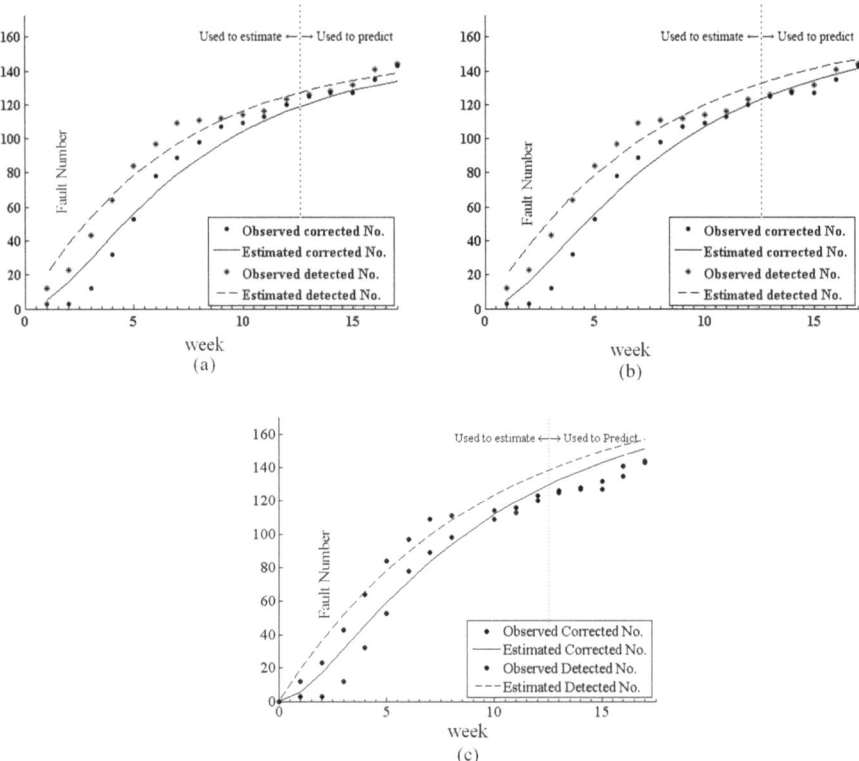

Fig. 5.4 Actual and estimate fault detection and correction processes. (**a**) Proposed model with WLS estimate, (**b**) ETD model with OLS estimate, (**c**) NHPP-based ETD model with WLS estimate ©[2016] John Wiley and Sons. Reprinted, with permission, from Liu et al. (2016)

5.4.4 Application to a New and Real Dataset with Unequal Time Interval

To illustrate more about the proposed model and WLS estimate, an original dataset collected in a real software project is applied in this part. In the software development of instrument and control systems for a nuclear power plant, some fault detection and correction datasets are collected. During the testing process, there are actually two groups participating at the same time. One debugging group detects the faults in current software via autotest (Ciupa et al. 2011) and submits them to the developing group who is assigned to correct faults. Thus, the collected data include the fault submitted time, which is used as fault detection data; fault-corrected time, which is used as fault correction data; and some additional properties such as fault severity and priority of correction process. A dataset named as Sys.A2 is chosen as the research object for this chapter.

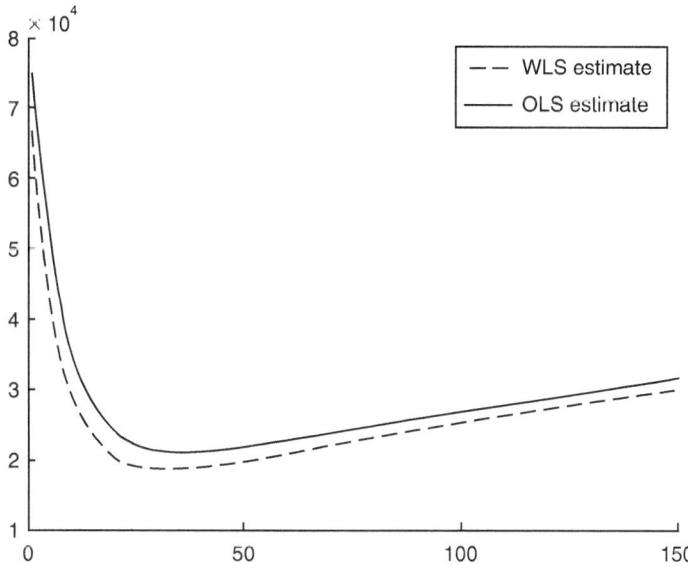

Fig. 5.5 Estimation of total cost ©[2016] John Wiley and Sons. Reprinted, with permission, from Liu et al. (2016)

Table 5.3 Cumulative detected fault number of Sys.A2 ©[2016] John Wiley and Sons. Reprinted, with permission, from Liu et al. (2016)

Test day	Detected no.	Test day	Detected no.	Test day	Detected no.
1	23	17	640	34	898
2	96	20	672	36	904
3	163	21	703	37	916
4	250	22	706	42	919
5	282	23	730	43	924
6	343	24	747	44	939
7	375	25	756	45	948
8	398	26	785	46	956
9	424	27	797	52	971
10	445	28	828	61	983
11	459	29	866	63	988
12	465	30	869	64	991
13	541	31	875	65	994
14	593	32	892	86	997
15	637	33	895	87	1000

The normalized cumulative fault number of Sys.A2 are listed in Tables 5.3 and 5.4. The testing days are counted with calendar time from which holidays and weekends are eliminated. Because the submission time is not equal to the fault-detected time and maybe delayed for random days, the sequence of test days is not continuous.

Table 5.4 Cumulative corrected fault number of Sys.A2 ©[2016] John Wiley and Sons. Reprinted, with permission, from Liu et al. (2016)

Test day	Corrected no.	Test day	Corrected no.	Test day	Corrected no.
1	6	18	523	36	811
2	15	19	529	38	837
3	17	23	532	39	849
4	20	24	535	40	855
5	23	25	541	41	860
6	35	26	581	42	863
7	44	27	608	43	869
10	55	28	622	44	884
11	84	29	631	45	927
12	105	30	660	46	942
13	128	31	689	65	945
14	163	32	712	68	959
15	203	33	744	69	962
16	302	34	773	70	965
17	483	35	794	74	968

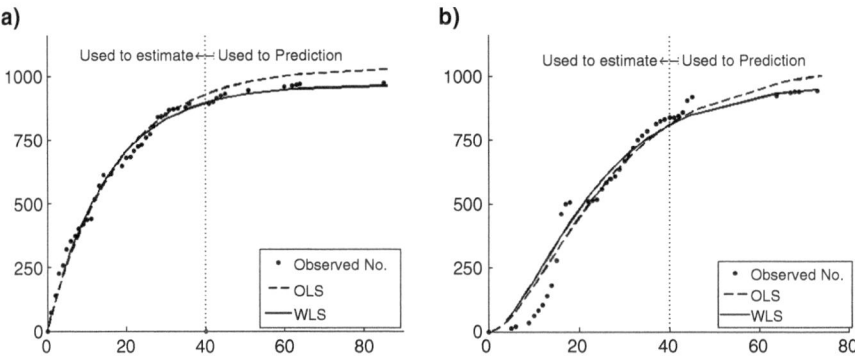

Fig. 5.6 Fault detection and correction with exponential time delay with OLS and WLS estimate (**a**) detection and (**b**) correction ©[2016] John Wiley and Sons. Reprinted, with permission, from Liu et al. (2016)

5.4.5 *Performance Comparison*

Due to the bad performance of model 3, this part focuses on the comparison of model 1 and model 2 with above datasets.

The first 40 days' test data of Sys.A2 is used to estimate the model parameters. Figure 5.6 shows the prediction curves with both of the OLS estimate (model 2) and WLS estimate (model 1). The estimated parameters and the fit goodness are shown in Table 5.5. The result demonstrates that the OLS is better than WLS in fitting observed dataset, but WLS has a better accuracy in prediction.

Table 5.5 Fitted results with exponential time delay ©[2016] John Wiley and Sons. Reprinted, with permission, from Liu et al. (2016)

Result		Estimates	
		Proposed model	Exponential time delay model
Parameter	N	967.63 ± 79.3450	1035.8424 ± 103.7698
	b	0.0652 ± 0.0116	0.0570 ± 0.0093
	μ	0.1091 ± 0.0359	0.0974 ± 0.0121
FDP	MSE of fitness	766.7523	924.5411
	MRE of prediction	0.0291	0.1744
FCP	MSE of fitness	5757.2243	4838.2014
	MRE of prediction	0.0756	0.1424

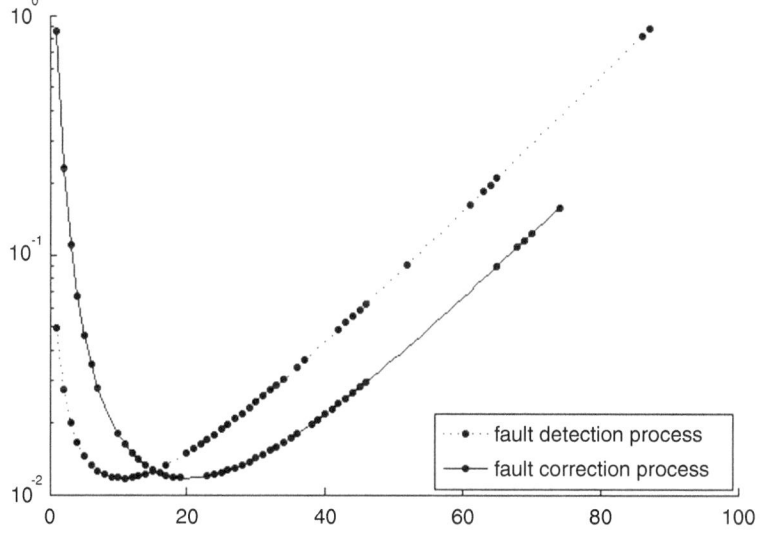

Fig. 5.7 Weight of WLS estimation ©[2016] John Wiley and Sons. Reprinted, with permission, from Liu et al. (2016)

In Table 5.5, the upper bound of N with WLSE is 1034.7626, which is lower than that of the LSE of N. Thus, in this case, the WLSE of N is significantly different from the LSE.

Figure 5.7 presents the weights of WLS, both the weights of FDP and FCP begin to increase with t, when $t_d \geq 11$ and $t_c \geq 17$.

The variable-term prediction results of WLS compared with OLS for FDP and FCP are shown in Fig. 5.8, respectively. A series of MRE from MRE (20) to MRE (70) are plotted against the end time of data used to estimate the model. Results illustrate that the WLS method has much lower prediction errors than the OLS method in both FDP and FCP, and the WLS method presents a good prediction in FDP with early data points, which is critical for making management scheme, such as prediction of release time and arrangement of testing effort.

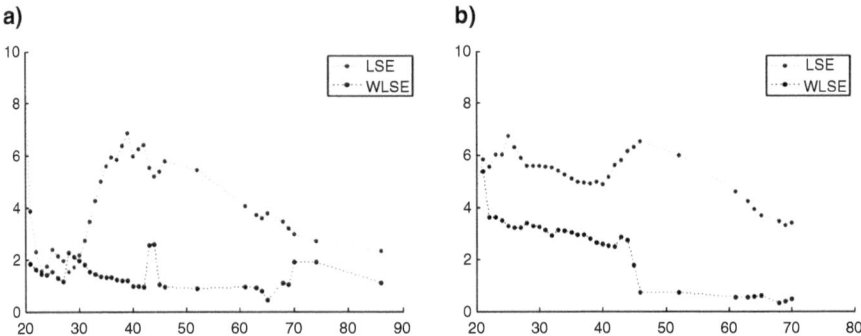

Fig. 5.8 Variable-term prediction ability of FDP (**a**) and FCP (**b**) ©[2016] John Wiley and Sons. Reprinted, with permission, from Liu et al. (2016)

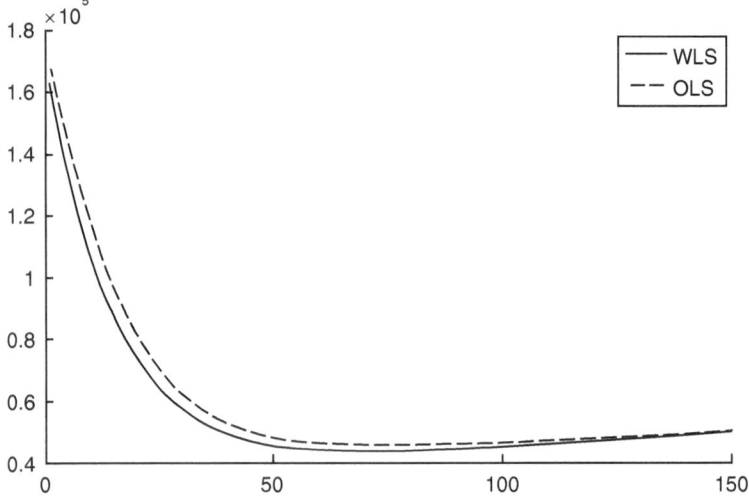

Fig. 5.9 Estimation of total cost ©[2016] John Wiley and Sons. Reprinted, with permission, from Liu et al. (2016)

5.4.6 Optimal Software Release Time Analysis

Figure 5.9 shows the plot of the total cost with the previous conditions, with comparison of different estimations. The optimal release times for the OLS estimation and WLS estimation model are 76 with a total cost of 45,088 and 70 with a total cost of 43,204, respectively. Considering the cost and to ensure high quality of the software, software managers should adopt WLS estimates as it is more stable and has better prediction performance (shown above). Also, the OLS estimation can be a reference to conservative estimates of software release time. These results are consistent with the former example.

5.5 Summary

In this chapter, a novel modeling framework for software FDP and FCP is developed, and its parameter estimation with WLS method is studied. The research is based on the hypothesis on GOS that the detection time and correction duration of each fault approximately and independently follow certain stochastic distribution, and the developed FDP and FCP models have a general framework. Using two datasets of which one is from published work and the other one is collected during the development of a practical software system, the application of the models is illustrated that the proposed model with the WLS estimate has lower prediction errors. This is because the proposed model contains the fault correction information as well as the detection information, and the WLS estimation weights more on the errors of recent data, which have more insights with the future test process.

There are some limitations in current research, such as it sets the exponential order statistic as example, and this model fails to fit the S-shaped fault detection dataset, which would be improved in future research. Moreover, other datasets should be applied and analyzed to support the research conclusions, and more data information, like severity and change point, should be considered to amend the software models.

References

Aktekin, T., & Caglar, T. (2013). Imperfect debugging in software reliability: A Bayesian approach. *European Journal of Operational Research, 227*, 112–121.

Chiu, K. C., Ho, J. W., & Huang, Y. S. (2009). Bayesian updating of optimal release time for software systems. *Software Quality Journal, 17*, 99–120.

Ciupa, I., Pretschner, A., Oriol, M., Leitner, A., & Meyer, B. (2011). On the number and nature of faults found by random testing. *Software Testing Verification & Reliability, 21*(1), 3–28.

Draper, N. R., & Smith, H. (1998). *Applied regression analysis, Wiley series in probability and statistics* (3rd ed.). New York: Wiley.

Dyer, M. (1991). Software reliability — What is it? *Software Testing, Verification and Reliability, 1*, 31–38.

Goel, A. L. (1985). Software reliability models: Assumptions, limitations, and applicability. *IEEE Transactions on Software Engineering, 11*, 1411–1423.

Gokhale, S. S., Lyu, M. R., & Trivedi, K. S. (2004). Analysis of software fault removal policies using a non-homogeneous continuous time Markov chain. *Software Quality Journal, 12*, 211–230.

Hu, Q. P., Xie, M., Ng, S. H., & Levitin, G. (2007). Robust recurrent neural network modeling for software fault detection and correction prediction. *Reliability Engineering & System Safety, 92*, 332–340.

Huang, C. Y., Kuo, C. S., & Luan, S. P. (2014). Evaluation and application of bounded generalized pareto analysis to fault distributions in open source software. *IEEE Transactions on Reliability, 63*, 309–319.

Huang, C. Y., & Hung, T. Y. (2010). Software reliability analysis and assessment using queueing models with multiple change-points. *Computers & Mathematics with Applications, 60*, 2015–2030.

Huang, C. Y., Lyu, M. R., & Kuo, S. Y. (2003). A unified scheme of some Nonhomogenous Poisson process models for software reliability estimation. *IEEE Transactions on Software Engineering, 29*, 261–269.

Ishii, T., Dohi, T., & Okamura, H. (2012). Software reliability prediction based on least squares estimation. *Quality Technology and Quantitative Management, 9*, 243–264.

Jelinski, Z., & Moranda, P. B. (1972). Software reliability research. In *Statistical computer performance evaluation* (pp. 465–484). New York: Academic Press.

Jukic, D. (2012). The L-p-norm estimation of the parameters for the Jelinski-Moranda model in software reliability. *International Journal of Computer Mathematics, 89*(4), 467–481.

Kapur, P. K., Pham, H., Anand, S., & Yadav, K. (2011). A unified approach for developing software reliability growth models in the presence of imperfect debugging and error generation. *IEEE Transactions on Reliability, 60*, 331–340.

Koh, S. L. P. (2008). Weighted least-square estimate for software error intensity. *Journal of the Chinese Institute of Industrial Engineers, 25*(2), 162–173.

Langberg, N., & Singpurwalla, N. D. (1985). A unification of some software-reliability models. *SIAM Journal on Scientific and Statistical Computing, 6*, 781–790.

Liu, Y., Li, D., Wang, L., & Hu, Q. (2016). A general modeling and analysis framework for software fault detection and correction process. *Software Testing, Verification & Reliability, 26*(5), 351–365.

Lyu, M. R. (1996). *Handbook of software reliability engineering*. Los Alamitos: IEEE computer Society Press.

Lyu, M. R. (2008). Software reliability engineering: A roadmap. *Future of Software Engineering, 7*, 153–170.

Malaiya, Y. K., Karunanithi, N., & Verma, P. (1992). Predictability of software-reliability models. *IEEE Transactions on Reliability, 41*(4), 539–546.

Miller, D. R. (1986). Exponential order statistic models of software reliability growth. *IEEE Transactions on Software Engineering, 12*(1), 12–24.

Musa, J. D., Iannino, A., & Okumoto, K. (1987). *Software reliability: Measurement, prediction, application*. New York: McGraw-Hill.

Ohishi, K., Okamura, H., & Dohi, T. (2009). Gompertz software reliability model: Estimation algorithm and empirical validation. *Journal of Systems and Software, 82*, 535–543.

Okamura, H., & Dohi, T., (2011). *Unification of software reliability models using Markovian arrival processes*. 2011 I.E. 17th Pacific Rim International Symposium on Dependable Computing (PRDC), 345 E 47TH ST, NEW YORK, NY 10017 USA, IEEE, 20–27.

Okamura, H., Dohi, T., & Osaki, S. (2013). Software reliability growth models with normal failure time distributions. *Reliability Engineering & System Safety, 116*, 135–141.

Peng, R., Li, Y. F., Zhang, W. J., & Hu, Q. P. (2014). Testing effort dependent software reliability model for imperfect debugging process considering both detection and correction. *Reliability Engineering & System Safety, 126*, 37–43.

Ross, S. M. (1985). Statistical estimation of software reliability. *IEEE Transactions on Software Engineering, 11*, 479–483.

Sahinoglu, M., & Can, Ü. (1997). Alternative parameter estimation methods for the compound Poisson software reliability model with clustered failure data. *Software Testing, Verification & Reliability, 7*(1), 35–57.

Schneidewind, N. (2009). Integrating testing with reliability. *Software Testing Verification & Reliability, 19*(3), 175–198.

Shibata, K., Rinsaka, K., Dohi, T., & Okamura, H., (2007). *Quantifying software maintainability based on a fault-detection/correction model*. 13th Pacific Rim International Symposium on Dependable Computing (PRDC 2007), IEEE, 35–42.

Wu, Y. P., Hu, Q. P., Xie, M., & Ng, S. H. (2008). Modeling and analysis of software fault detection and correction process by considering time dependency. *IEEE Transactions on Reliability, 56*, 629–642.

Xie, M. (1993). Software reliability models—A selected annotated bibliography. *Software Testing, Verification & Reliability, 3*(1), 3–28.

Xie, M., Hu, Q., Wu, Y., & Ng, S. (2007). A study of the modeling and analysis of software fault-detection and fault-correction processes. *Quality and Reliability Engineering International, 56*, 459–470.

Yamada, S., & Fujiwara, T. (2001). Testing-domain dependent software reliability growth models and their comparisons of goodness-of-fit. *International Journal of Reliability Quality and Safety Engineering, 8*, 205–218.

Chapter 6
Reliability of Multi-release Open-Source Software

Previous chapters studied the fault detection process and correction process for the software with single release. In this chapter, a fault removal modeling framework for software reliability with semigrouped data is studied and extended into multi-released software. Also, the corresponding parameter estimation is carried out with maximum likelihood estimation method. One test dataset with three releases from a practical software project is applied with the proposed framework, which shows satisfactory performance with the results.

6.1   Introduction

Most parametric SRGMs regard the software test as a black box testing and utilize the fault detection record to describe the stochastic behavior of software failure in fault detection process (FDP). However, most of them are established under the idealized assumption that all the faults are removed instantaneously and perfectly. Some researchers also took into consideration the fault correction process (FCP) and used the fault correction data to estimate the correction time delay (Schneidewind 2001; Xie et al. 2007; Wu et al. 2008; Hu et al. 2007; Peng et al. 2014; Huang and Huang 2008, 2010; Gaver and Jacobs 2014; Lin 2011; Lin and Li 2014). All of those models are conducted with the grouped fault data, while in practical, software test records often include more detailed information, such as the severity of each fault, the rough calendar time when one fault is detected and corrected, etc. It is always reasonable to use the grouped data to modeling the FDP along due to the uncertainty of fault-detected time. When the FCP is involved, instead of using the cumulated corrected faults' number in each time interval separately as earlier studies, the exact removal time interval of each fault can be taken into modeling. This type of data is named as the semigrouped data. Since the semigrouped data specify the one-to-one correspondence with FCP and FDP, it can provide more stability in estimation and improve the accuracy of models' prediction.

R. Peng et al., *Software Fault Detection and Correction: Modeling and Applications*, SpringerBriefs in Computer Science, https://doi.org/10.1007/978-981-13-1162-8_6

In this study, we establish a parametric modeling framework, using the semigrouped dataset which includes the time interval when a fault is detected and corrected. The rest of this chapter is organized as follows. Section 6.2 outlines the framework of modeling FDP and FCP with semigrouped dataset. In Sect. 6.3 the parameter estimation with ML estimate is developed, and both the point estimation and interval estimation are studied. Section 6.4 gives an example to apply the proposed framework with an actual test dataset.

6.2   Modeling Framework in FDP and FCP with Semigrouped Test Data

6.2.1   Fault Correction Process Modeling

Detailed information often brings more accuracy into modeling and improves its prediction. The test record actually contains many details about testing; however, only limited things are extracted to conduct models. Thus, to incorporate more information, the detecting time and correcting time of each fault are used in this new approach.

6.2.2   Fault Removing Matrix

To describe the detected and corrected fault count in time sequences and illustrate the relationship between them, the fault removing matrix (FRM) is established.

Figure 6.1 shows an example of FRM, in which the number of faults detected in the same time interval $(t_i, i = 0, 1, 2,\ldots,W)$ is listed in each column of FRM and separated by different time intervals which refer to their corrected time $(t_j, j = 0, 1, 2,\ldots, W)$. Thus, the sum of each row is the total detected fault count in each time interval, and the sum of each column is the total corrected fault count in each time interval. In some situation, not all the faults detected in same time interval can be removed at the end time of test. To describe those detected but uncorrected faults (DUCF), a row of uncorrected fault count is added at the end of the matrix as shown in Fig. 6.1.

Fig. 6.1 Fault removing matrix ©[2015] IEEE (Reprinted with permission, from Liu et al. 2015)

		FDP		
	Week	1	2	3
FCP	1	2		
	2	1	1	
	3	0	3	2
DUCF		1	0	1

6.2.3   Modeling FDP and FCP Based on the Fault Removing Matrix

Theorem 1 Given that the fault detection process $\{N_d(t), \ t > 0\}$, the number of faults which are detected between $(t_{i-1}, \ t_i, \ i = 1, 2, \ldots, W)$ and corrected between $(t_{j-1}, \ t_j, j = i-1, i, \ldots, W)$, denoted as $\Delta N_c(t_{i-1}, t_i, t_{j-1}, t_j)$, is an independent Poisson distribution with the mean value function

$$
\begin{aligned}
&E\left[\Delta N_c\left(t_{i-1}, t_i, t_{j-1}, t_j\right)\right] \\
&= \int_{t_{i-1}}^{t_i} \left[G\left(t_j - x\right) - G\left(t_{j-1} - x\right)\right]\lambda_d(x)dx
\end{aligned}
\tag{6.1}
$$

in which $G(t) = 0$ when $t < 0$.

Proof The probability that a fault is detected at t_f, between $(t_{i-1}, \ t_i)$ is

$$
P\{t_f|t_{i-1} \le t_f < t_i\} = \frac{\lambda_d\left(t_f\right)}{m_d(t_i) - m_d(t_{i-1})}
\tag{6.2}
$$

Then, the probability of the fault which is detected between $(t_{i-1}, \ t_i)$ and corrected in $(t_{j-1}, \ t_j)$ is

$$
\begin{aligned}
P_{i|j} &= \int_{t_{i-1}}^{t_i} G\left(t_j - x\right) \frac{\lambda_d(x)}{m_d(t_i) - m_d(t_{i-1})} dx \\
&\quad - \int_{t_{i-1}}^{t_i} G\left(t_{j-1} - x\right) \frac{\lambda_d(x)}{m_d(t_{i-1}) - m_d(t_{i-1})} dx \\
&= \left[m_d(t_i) - m_d(t_{i-1})\right]^{-1} \\
&\quad \cdot \int_{t_{i-1}}^{t_i} \left[G\left(t_j - x\right) - G\left(t_{j-1} - x\right)\right]\lambda_d(x)dx
\end{aligned}
\tag{6.3}
$$

It can be seen that

$$
\begin{aligned}
\lim_{W\to\infty} &\sum_{j=i}^{W} P_{i|j} \\
&= \left[m_d(t_i) - m_d(t_{i-1})\right]^{-1} \\
&\quad \cdot \int_{t_{i-1}}^{t_i} \left[G(t_\infty - x) - G(t_{i-1} - x)\right]\lambda_d(x)dx \\
&= 1
\end{aligned}
\tag{6.4}
$$

The probability that a fault is detected between $(t_{i-1}, \ t_i)$ and has not corrected till time t_W is

$$q_{i|W} = 1 - \sum_{j=i}^{W} p_{i|j} \tag{6.5}$$

Since the FCP is assumed to be NHPP, then

$$
\begin{aligned}
P&\left\{ \bigcup_{j=i}^{W} \Delta N_c(t_i, t_{i+1}, t_j, t_{j+1}) = \Delta n_{i|j}, \mathrm{DUCF}_i = \Delta n_{i|\infty} \right\} \\
&= P\left\{ \bigcup_{j=i}^{W} \Delta N_c(t_i, t_{i+1}, t_j, t_{j+1}) = \Delta n_{i|j}, \mathrm{DUCF}_i = \Delta n_{i|\infty} \Big| N_d(t_i, t_{i+1}) = n_i \right\} \\
&\quad \times P\{N_d(t_i, t_{i+1}) = n_i\} \\
&= n_i! \prod_{j=i}^{W} \left(p_{i|j}^{n_{i|j}} / \Delta n_{i|j}! \right) q_{i|W}^{\Delta n_{i|\infty}} / \Delta n_{i|\infty}! \\
&\quad \times \frac{[m_d(t_i) - m_d(t_{i-1})]^{n_i}}{n_i!} \exp[-(m_d(t_i) - m_d(t_{i-1}))] \\
&= \prod_{j=i}^{W} \left(\left[(m_d(t_i) - m_d(t_{i-1})) p_{i|j} \right]^{\Delta n_{i|j}} / \Delta n_{i|j}! \right) \\
&\quad \times \exp\left[-(m_d(t_i) - m_d(t_{i-1})) p_{i|j} \right] \\
&\quad \times \left(\left[(m_d(t_i) - m_d(t_{i-1})) q_{i|W} \right]^{\Delta n_{i|\infty}} / \Delta n_{i|\infty}! \right) \\
&\quad \times \exp\left[-(m_d(t_i) - m_d(t_{i-1})) q_{i|W} \right]
\end{aligned}
\tag{6.6}
$$

in which, DUCF_i is the count of faults which are detected between (t_{i-1}, t_i) but not corrected until t_W and

$$\Delta n_{i|\infty} = n_i - \sum_{j=i}^{W} \Delta n_{i|j}$$

The marginal distribution of $\Delta N_c(t_{i-1}, t_i, t_{j-1}, t_j)$ can be derived as

$$
\begin{aligned}
P\{\Delta N_c(t_i, t_{i+1}, t_j, t_{j+1}) = \Delta n_{ij}\} = \\
\frac{\left[(m_d(t_i) - m_d(t_{i-1})) p_{i|j} \right]^{\Delta n_{i|j}}}{\Delta n_{i|j}!} \exp\left[-(m_d(t_i) - m_d(t_{i-1})) p_{i|j} \right]
\end{aligned}
\tag{6.7}
$$

Then, because of $N_c(t_{i-1}, t_i, t_{j-1}, t_j)$ is independent with each other, the number of faults which are detected between (t_{i-1}, t_i) and corrected between (t_{j-1}, t_j) is an independent Poisson distribution with the mean value function (Ross 1996),

$$E\left[\Delta N_c\left(t_{i-1},t_i,t_{j-1},t_j\right)\right] = \int_{t_{i-1}}^{t_i} \left[G\left(t_j - x\right) - G\left(t_{j-1} - x\right)\right]\lambda_d(x)\mathrm{d}x$$

It can be inferred that

$$\begin{cases} \displaystyle\lim_{W\to\infty} \sum_{j=i}^{W}\Delta N_c\left(t_{i-1},t_i,t_{j-1},t_j\right) = m_d(t_i) - m_d(t_{i-1}) \\[2ex] \displaystyle\sum_{i=1}^{j}\Delta N_c\left(t_{i-1},t_i,t_{j-1},t_j\right) = m_c\left(t_j\right) - m_c\left(t_{j-1}\right) \end{cases} \tag{6.8}$$

Based on above analysis, the FRM and expected FRM (EFRM) can be defined with (6.1)

$$EFRM_{ji} = E\left[\Delta N_c\left(t_{i-1},t_i,t_{j-1},t_j\right)\right]$$
$$EFRM_{(W+1)i} = [m_d(t_i) - m_d(t_{i-1})]q_{i|W}$$

Thus, the sum of each row of EFRM is the expected detected fault count in each time interval, and the sum of each column is the expected corrected fault count in each time interval. Then, based on the (6.6) and the independent assumption of FDP, the joint probability is

$$\begin{aligned} P&\left\{\bigcup_{i=1}^{W}\left[\bigcup_{j=i}^{W}\Delta N_c\left(t_i,t_{i+1},t_j,t_{j+1}\right) = n_{i|j}, \text{DUCF}_i = n_{i|\infty}\right]\right\} \\ &= \prod_{i=1}^{W}P\left\{\bigcup_{j=i}^{W}\Delta N_c\left(t_i,t_{i+1},t_j,t_{j+1}\right) = n_{i|j}, \text{DUCF}_i = n_{i|\infty}\right\} \\ &= \prod_{i=1}^{W}\left\{\prod_{j=i}^{W}\left(\left[E\left(\Delta N_c\left(t_{i-1},t_i,t_{j-1},t_j\right)\right)\right]^{n_{i|j}} / n_{i|j}!\right)\right. \\ &\qquad\qquad \times \exp\left[-E\left(\Delta N_c\left(t_{i-1},t_i,t_{j-1},t_j\right)\right)\right] \\ &\qquad\qquad \times \left(\left[(m_d(t_i) - m_d(t_{i-1}))q_{i|W}\right]^{n_{i|\infty}} / n_{i|\infty}!\right) \\ &\qquad\qquad \left.\times \exp\left[-(m_d(t_i) - m_d(t_{i-1}))q_{i|W}\right]\right\} \end{aligned} \tag{6.9}$$

With the definition of FRM and EFRM, a concise joint probability equation can be written as

$$P\{FRM\} = \prod_{j=1}^{W+1} \prod_{i=1}^{\min(W,j)} \frac{EFRM_{ji}^{FRM_{ji}}}{FRM_{ji}!}\exp\left(-EFRM_{ji}\right)$$

6.2.4 Extended Modeling Framework with Multiple Releases

In multi-release circumstances, when new functions or patches are added to the software, new faults are also embedded into the new version. As a result, the removed faults in the new version can be decomposed into two groups: newly detected faults and the detected but uncorrected faults (DUCFs) in the previous versions. Thus, the expected number of DUCFs which are discovered between time interval $(t_i^{(v_1)}, t_{i+1}^{(v_1)})$ in the version v_1 and corrected between time interval $(t_j^{(v_2)}, t_{j+1}^{(v_2)})$ in the version v_2 is given by

$$
\begin{aligned}
E\left(n_{ij}^{(v_1)(v_2)}\right) \\
= \left(m_{dv_1}\left(t_i^{(v_1)}\right) - m_{dv_1}\left(t_{i-1}^{(v_1)}\right)\right) q_{i|W_{v_1}}^{(v_1)} \cdot G_{v_2}\left(t_j^{(v_2)} - t_{j-1}^{(v_2)}\right) \\
\cdot \begin{cases} \prod_{k=v_1+1}^{v_2-1} [1 - G_k(T_k)], & v_2 \geq v_1 + 2 \\ 1 & , \quad v_2 = v_1 + 1 \end{cases}
\end{aligned} \tag{6.10}
$$

in which $m_{dv1}(t)$ refers to the MVF of FDP in version v_1, $G_{v2}(t)$ refers to the CDF of correcting time delay in version v_2, and T_k is the end time of test on version k.

The expected number of faults which are detected in version v_1 and have not corrected till T_V can be derived as

$$
E\left(n_{i\infty}^{(v_1)(V)}\right) = \left(m_{dv_1}\left(t_i^{(v_1)}\right) - m_{dv_1}\left(t_{i-1}^{(v_1)}\right)\right) q_{i|W_{v_1}}^{(v_1)} \cdot \prod_{k=v_1+1}^{V} [1 - G_k(T_k)] \tag{6.11}
$$

Then, the joint probability with V releases can be written as

$$
\begin{aligned}
P\{D_1, D_2, \ldots, D_V\} \\
= \prod_{k=1}^{V} \prod_{j=1}^{W_k+1} \prod_{i=1}^{\min(W_k,j)} \frac{\left[EFRM_{ji}^{(k)}\right]^{FRM_{ji}^{(k)}}}{FRM_{ji}^{(k)}!} \exp\left(-EFRM_{ji}^{(k)}\right) \\
\cdot \prod_{v_1=1}^{V-1} \prod_{v_2=v_1+1}^{V} \prod_{j=1}^{T_{v2}} \prod_{i=1}^{T_{v1}} \frac{E\left(n_{ij}^{(v_1)(v_2)}\right)^{n_{ij}^{(v_1)(v_2)}}}{n_{ij}^{(v_1)(v_2)}!} \exp\left(-E\left(n_{ij}^{(v_1)(v_2)}\right)\right) \\
\cdot \prod_{v_1=1}^{V-1} \prod_{i=1}^{T_{v1}} \frac{E\left(n_{i\infty}^{(v_1)(V)}\right)^{n_{i\infty}^{(v_1)(V)}}}{n_{i\infty}^{(v_1)(V)}!} \exp\left(-E\left(n_{i\infty}^{(v_1)(V)}\right)\right)
\end{aligned}
$$

$$
\tag{6.12}
$$

Considering the impact of multiple releases, we can rewrite the MVF of FCP as

$$m_c^{(v_2)}(t_j) = m_{cv_2}(t_j) + \sum_{v_1=1}^{v_2-1}\sum_{i=1}^{T_{v_1}} E\left(n_{ij}^{(v_1)(v_2)}\right)$$

$$= m_{cv_2}(t_j) + \sum_{v_1=1}^{v_2-1}[m_{dv_1}(T_{v_1}) - m_{cv_1}(T_{v_1})] \cdot G_{v_2}(t_j) \cdot \begin{cases} \prod_{k=v_1+1}^{v_2-1}[1 - G_k(T_k)], & v_2 \geq v_1 + 2 \\ 1 & , & v_2 = v_1 + 1 \end{cases}$$

$$(6.13)$$

6.3   Parameter Estimation and Comparison Criteria

In this section, the maximum likelihood method with a new likelihood function is conducted based on the fault removing matrix. For simplification, this section just set the single released software as an example. The parameter estimation with multiple-release models can be derived similarly and will not be covered here.

6.3.1   Point Estimation

A general form of the likelihood function for the combined FDP and FCP can be obtained with (6.9). The log form of joint probability equation is used here

$$L(\theta|FRM) = \sum_{j=1}^{W+1}\sum_{i=1}^{\min(W,j)} \left(FRM_{ji} \cdot \ln EFRM_{ji} - \ln\left(FRM_{ji}!\right) - EFRM_{ji}\right) \quad (6.14)$$

in which θ denotes the vector of parameters in the FDP and FCP models. The optimal solution can be obtained by utilizing iterative methods such as Newton's method, quasi-Newton's method, Fisher's scoring method, and Nelder-Mead method.

For comparison, the common-used least square estimation is also applied in estimation. The parameters' estimates are the optimal solution of the following nonlinear equation (Xie et al. 2007):

$$\min_{\theta} S(\theta) = \sum_{i=1}^{W}\left(m_d(t_i, \theta) - n_i\right)^2 + \sum_{j=1}^{W}\left(m_c(t_j, \theta) - m_j\right)^2 \quad (6.15)$$

in which n_i and m_j refer to the observed cumulative fault counts in FDP and FCP by testing time t_i/t_j. Above problem can be solved with the build-in function of MATLAB.

6.3.2   Interval Estimation

Confidence interval can present the credibility of estimation. The Fisher information matrix (Lawless 2011) provides an approach for MLE to calculate the asymptotic covariance matrix.

If $\theta \in R^P$, the Fisher information matrix is

$$H = \left[E\left(-\frac{\partial^2 \ln(L(\theta))}{\partial\theta_i \partial\theta_j} \right) \right]_{P \times P} \quad i = 1, 2, \ldots, P \; j = 1, 2, \ldots, P \qquad (6.16)$$

Then, the asymptotic covariance matrix V of the ML estimates for parameters is the inverse of the Fisher information matrix shown in (6.12)

$$V = H^{-1} \qquad (6.17)$$

The two-sided approximate $100\alpha\%$ confidence limits for parameter θ can be obtained as

$$\widehat{\theta} = \widehat{\theta} \pm Z_\alpha \sqrt{diag(V)} \qquad (6.18)$$

where Z_α is the α-quantile of the standard normal distribution.

The confidence interval of the faults expected counts could be obtained as

$$m(t) \pm Z_\alpha \sqrt{V_0 \times V \times V_0^T} \qquad (6.19)$$

in which

$$V_0 = \left[\frac{\partial m(t_0)}{\partial\theta} \right]_{1 \times P} \qquad (6.20)$$

The confidence intervals for LSE method can be found in [26] and will not be covered here.

6.3.3 Comparison Criteria

The mean squared error (MSE) and adjusted R-square (R^2) are used to compare the models' goodness of fit:

$$\begin{cases} MSE_k^{(d)} = \dfrac{1}{k}\sum_{i=1}^{k}\left(m_d(t_i,\theta) - n_i\right)^2 \\[4mm] MSE_k^{(d)} = \dfrac{1}{k}\sum_{j=1}^{k}\left(m_c(t_j,\theta) - m_j\right)^2 \end{cases} \tag{6.21}$$

$$\begin{cases} R_d^2 = 1 - \dfrac{\sum\limits_{i=1}^{k}\left(m_d(t_i,\theta) - n_i\right)^2}{\sum\limits_{i=1}^{k}\left(n_i - \bar{n}_k\right)^2}, \bar{n}_k = \dfrac{1}{k}\sum\limits_{i=1}^{k}n_i \\[8mm] R_c^2 = 1 - \dfrac{\sum\limits_{j=1}^{k}\left(m_c(t_j,\theta) - m_j\right)^2}{\sum\limits_{j=1}^{k}\left(m_j - \bar{m}_k\right)^2}, \bar{m}_k = \dfrac{1}{k}\sum\limits_{i=1}^{k}m_j \end{cases} \tag{6.22}$$

When new parameters are added to the model, the R^2 increases and may lead to over-fitting. Thus, it is reasonable to use the adjusted R^2 instead of R^2, as shown below.

$$\begin{cases} adj\_R_d^2 = 1 - \dfrac{(k-1)\sum\limits_{i=1}^{k}\left(m_d(t_i,\theta) - n_i\right)^2}{(k-1-p)\sum\limits_{i=1}^{k}\left(n_i - \bar{n}_k\right)^2}, \bar{n}_k = \dfrac{1}{k}\sum\limits_{i=1}^{k}n_i \\[8mm] adj\_R_c^2 = 1 - \dfrac{(k-1)\sum\limits_{j=1}^{k}\left(m_c(t_j,\theta) - m_j\right)^2}{(k-1-p)\sum\limits_{j=1}^{k}\left(m_j - \bar{m}_k\right)^2}, \bar{m}_k = \dfrac{1}{k}\sum\limits_{i=1}^{k}m_j \end{cases} \tag{6.23}$$

where p is the number of model parameters.

To study the prediction accuracies of this model with different estimation methods, the one-step forward prediction is used to quantify this ability (Malaiya 1992). This approach predicts the detected fault number and the corrected fault number in next time interval, using the former partial dataset. The relative errors (RE) are used to measure the predicting errors, which is

$$\begin{cases} RE_d = \dfrac{m_d\left(t_{l+1}, \widehat{\theta}\right) - n_{l+1}}{n_{l+1}} \\ RE_c = \dfrac{m_c\left(t_{l+1}, \widehat{\theta}\right) - m_{l+1}}{m_{l+1}} \end{cases} \tag{6.24}$$

where $\widehat{\theta}$ is estimated from the partial dataset (n_i, m_i, t_i, $i = 1, 2, 3. . .,l$).

6.4 Illustrative Example

6.4.1 Dataset Description

To illustrate the proposed approach, a software test dataset is collected from an online bug tracking system, *Mozilla Bugzilla*, which was developed and applied by Mozilla foundation (*https://bugzilla.mozilla.org/*). Many organizations and enterprises especially the open-source software utilize this system to manage their software developing process, such as the core development of Linux, Apache, and GNOME.

We collected and reorganized the fault tracking data on Firefox 3.0, 3.5, and 3.6 from Bugzilla, which are three successive versions as shown in Tables 6.1, 6.2, and 6.3. It can be seen that at the end of test on the Firefox 3.0, 3.5, 3.6, there are 17, 15, and 9 faults, respectively, which have been detected without correction. Compare to the total number of detected faults, the numbers of DUCFs account for 26.15%, 29.41%, and 31.03%, respectively, which are quite large than one's thought.

6.4.2 Model Application with Single Released Software

Model Estimation

In this part, the proposed models are applied with the collected dataset on Firefox 3.0. Using the LSE and MLE, the estimates can be obtained, given in Table 6.4. Both of the exponential and Weibull time delay models are trained.

From the results shown in Table 6.4, the models estimated with LSE have overall lower error than MLE due to the smaller MSE, while one can see from the Figs. 6.2 and 6.3 that the MLE does better than LSE in fitting the later data points. The Weibull time delay model has much better goodness of fit than the exponential time delay model especially in the data of FCP. From the point of view of R^2 and adjusted R^2, the estimation of Weibull time delay model with MLE has the biggest correlation with the observed data. Comparing the parameters in these models, we

Table 6.1 Faults test record of Firefox 3.0 ©[2015] IEEE

Fault no.	Detected week	Corrected week	Fault no.	Detected week	Corrected week
1	1	91	34	9	10
2	1	1	35	10	44
3	1	4	36	11	11
4	1	17	37	13	13
5	1	–	38	13	14
6	1	1	39	13	44
7	1	92	40	15	35
8	1	1	41	18	19
9	1	31	42	19	19
10	2	43	43	20	36
11	2	5	44	22	30
12	2	3	45	23	48
13	3	4	46	25	26
14	3	–	47	26	68
15	3	92	48	28	31
16	3	13	49	28	28
17	4	84	50	29	31
18	4	33	51	33	63
19	4	9	52	34	68
20	4	10	53	35	50
21	4	45	54	36	36
22	4	4	55	37	61
23	4	5	56	41	57
24	4	83	57	42	–
25	4	6	58	42	45
26	5	9	59	42	48
27	5	18	60	43	52
28	6	6	61	46	96
29	6	6	62	47	47
30	8	31	63	52	53
31	8	8	64	52	73
32	8	61	65	53	66
33	9	12	66	54	59

Reprinted, with permission, from Liu et al. (2015)

can see that the estimates of γ and a are fairly equal, no matter what kind of time delay or estimation method is used.

Figures 6.2 and 6.3 also show that the confidence intervals derived from the LSE method appear to be very small. The confidence intervals provided by the MLE method seem to be more realistic because they are wider and contain most observed points. Therefore, the estimation based on MLE can be more reasonable especially when conservative approximation is needed, while both of the

Table 6.2 Fault test record of Firefox 3.5 ©[2015] IEEE

Fault no.	Detected week	Corrected week	Fault no.	Detected week	Corrected week
67	55	55	93	62	65
68	55	–	94	62	85
69	55	55	95	65	70
70	55	55	96	65	70
71	55	92	97	65	73
72	55	62	98	65	108
73	55	72	99	65	68
74	56	75	100	66	66
75	56	–	101	66	66
76	56	65	102	66	96
77	57	57	103	67	77
78	57	84	104	67	75
79	57	57	105	68	69
80	57	–	106	70	70
81	57	–	107	72	–
82	58	61	108	73	75
83	58	58	109	75	76
84	59	59	110	76	77
85	59	–	111	76	78
86	59	60	112	76	76
87	59	60	113	77	77
88	60	61	114	79	79
89	61	65	115	79	89
90	61	–	116	81	84
91	61	64	117	82	83
92	61	65	118	83	–
			119	83	110

Reprinted, with permission, from Liu et al. (2015)

estimations cannot fit the early part of the FCP well, which may result from changes in test effort and the difficulties in correction.

Model Prediction

In this part, the one-step forward prediction is applied. Using the parameters got from LSE and MLE, the expected fault counts in next time interval is computed and compared with observed value. Both of the exponential and Weibull time delay models are trained.

Figure 6.4 depicts the relative errors of above models against the test time, reflecting the prediction capacity of each model. It seems that Weibull time delay model with LSE tends to be biased to the underestimation side in predicting both FDP

Table 6.3 Fault test record of Firefox 3.6 ©[2015] IEEE

Fault no.	Detected week	Corrected week	Fault no.	Detected week	Corrected week
120	85	102	135	95	95
121	86	86	136	99	99
122	86	130	137	99	106
123	89	90	138	99	100
124	89	–	139	111	–
125	90	93	140	111	115
126	93	–	141	114	–
127	93	95	142	115	119
128	94	–	143	115	–
129	94	115	144	119	–
130	94	122	145	120	120
131	94	100	146	120	–
132	95	95	147	124	–
133	95	129	148	124	124
134	95	97	149	131	–

Reprinted, with permission, from Liu et al. (2015)

Table 6.4 Goodness of fit results of different models for the single released dataset ©[2015] IEEE

Model	Exponential time delay	
Method	MLE	LSE
Estimates	$a = 71.5361 \pm 14.4885$,	$a = 57.2758 \pm 2.2018$,
	$\gamma = 0.0474 \pm 0.0011$,	$\gamma = 0.0862 \pm 0.0120$,
	$\mu = 0.0505 \pm 0.0114$	$\mu = 0.0490 \pm 0.0052$
MSE_d	27.3606	17.9104
MSE_c	37.0230	22.5446
R_d^2	0.9806	0.9527
R_c^2	0.9765	0.9786
adj_ R_d^2	0.9795	0.9499
adj_ R_c^2	0.9751	0.9773
Model	Weibull time delay	
Method	MLE	LSE
Estimates	$a = 71.5887 \pm 10.8283$,	$a = 60.7374 \pm 2.1912$,
	$\gamma = 0.0472 \pm 0.0055$,	$\gamma = 0.0706 \pm 0.0079$,
	$\lambda = 16.5491 \pm 7.0052$,	$\lambda = 18.1968 \pm 3.2701$,
	$k = 0.4199 \pm 0.0980$	$k = 0.3715 \pm 0.1136$
MSE_d	27.3941	16.1856
MSE_c	10.7307	6.6705
R_d^2	0.9807	0.9666
R_c^2	0.9849	0.9789
adj_ R_d^2	0.9795	0.9646
adj_ R_c^2	0.9840	0.9776

Reprinted, with permission, from Liu et al. (2015)

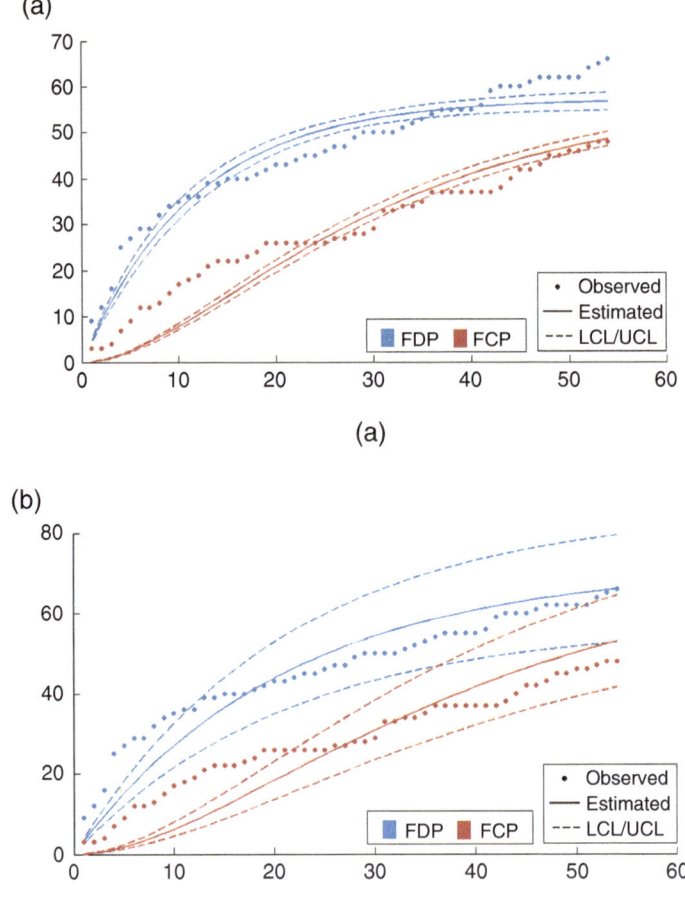

Fig. 6.2 Proposed model with exponential time delay ©[2015] IEEE. (**a**) LSE. (**b**) MLE. (Reprinted, with permission, from Liu et al. 2015)

and FCP after testing for 20 weeks. The exponential time delay model with LSE tends to have more underestimation in predicting FDP and tends to have overestimation in predicting FCP at early time of testing, while the exponential time delay mode with MLE seems to overestimate the fault corrected count all the time. The model, which has the least relative error in predicting FDP and FCP, is Weibull time delay model with MLE.

The above comparison shows that the time delay model with MLE can provide a better and more stable result in fitting the latest dataset and predicting future process.

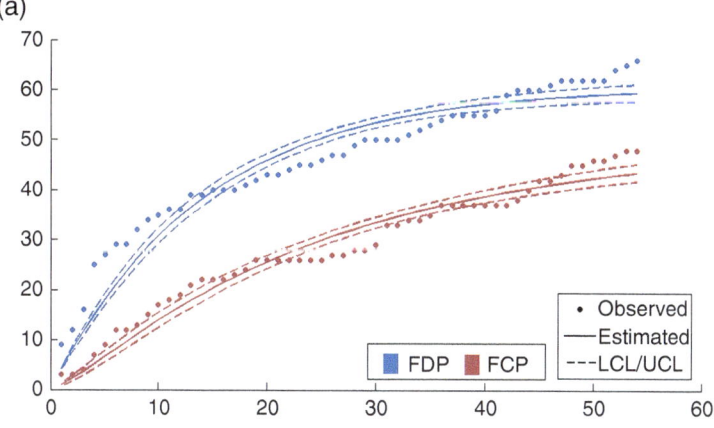

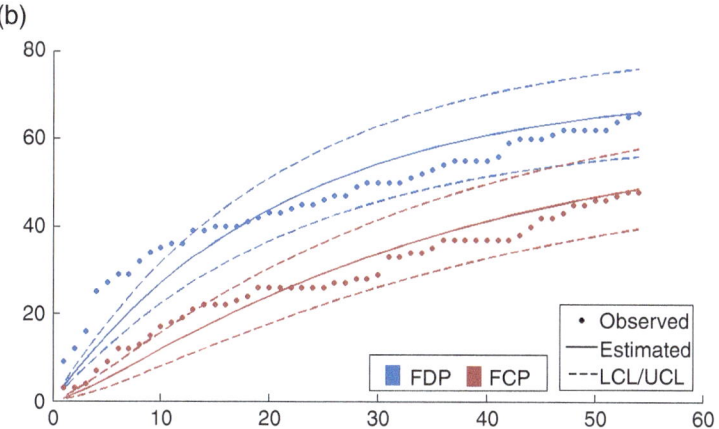

Fig. 6.3 Proposed model with Weibull time delay ©[2015] IEEE. (**a**) LSE. (**b**) MLE. (Reprinted, with permission, from Liu et al. 2015)

6.4.3   Model Application with Multiple Released Software

Model Estimation

In this subsection, our proposed models are tested with the aforementioned three-version dataset. We obtain the estimates listed in Table 6.5. Both the exponential time delay model and Weibull time delay model are trained. The expected numbers of faults are plotted in Fig. 6.5 against test time.

Both the MSE and R^2 criteria in the foregoing results show that the models with Weibull time delay display a much better goodness of fit than the exponential time

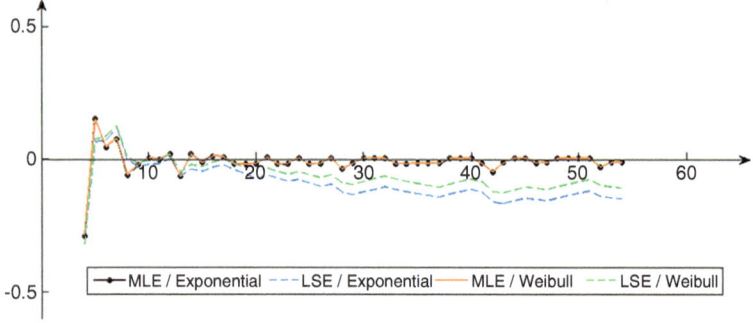

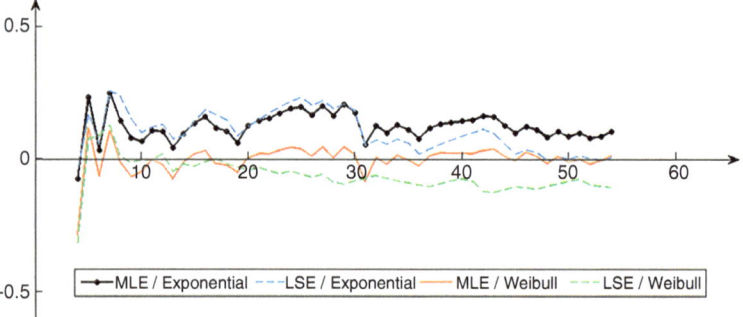

Fig. 6.4 Relative error of predicted md(t) and mc(t) of proposed models © [2015] IEEE. (Reprinted, with permission, from Liu et al. 2015)

delay models in terms of both FDP and FCP. Comparing the parameters in these models, we can see that the estimates of γ_i and a_i are fairly equal, no matter what kind of time delay is used.

From Fig. 6.5, we can see that the estimated curve is more closed to the observed curve at the later part of the data. Influenced by the correction of the DUCFs, the estimated FCP curve with Weibull time delay appears to have a little jump at the beginning of the version II and III.

Model Prediction

In this part, we still use the one-step forward prediction to compare the predictive capacity of the models with exponential or Weibull time delay.

Figure 6.6 depicts the relative errors of the above models against the test time. It seems that both of the models can give similar prediction on the number of detected faults in next week, while the model with Weibull time delay has a much better accuracy on predicting FCP, compared to the model with exponential time

Table 6.5 Goodness of fit results of different models for the three-version dataset ©[2015] IEEE

Model	$\Delta t = \text{Exp}(\mu)$					
	Multi-release					
Version	I		II		III	
Estimates	$a_1 = 71.5361$		$a_2 = 77.9775$		$a_3 = 46.7744$	
	$r_1 = 0.0474$		$r_2 = 0.0393$		$r_3 = 0.0214$	
	$\mu_1 = 0.0505$		$\mu_2 = 0.0900$		$\mu_3 = 0.0525$	
Comparing criteria	MSE_d	16.8479	R_d^2	0.9956	$adj\_R_d^2$	0.9955
	MSE_c	19.5717	R_c^2	0.9952	$adj\_R_c^2$	0.9951
Model	$\Delta t = \text{Weibull } (k, \lambda)$					
	Multi-release					
Version	I		II		III	
Estimates	$a_1 = 71.5904$		$a_2 = 62.0679$		$a_3 = 40.4872$	
	$r_1 = 0.0472$		$r_2 = 0.0663$		$r_3 = 0.0281$	
	$k_1 = 0.4205$		$k_2 = 0.4377$		$k_3 = 0.4186$	
	$\lambda_1 = 16.5593$		$\lambda_2 = 16.9357$		$\lambda_3 = 38.7094$	
Comparing criteria	MSE_d	13.5908	R_d^2	0.9964	$adj\_R_d^2$	0.9963
	MSE_c	7.5591	R_c^2	0.9978	$adj\_R_c^2$	0.9977

Reprinted, with permission, from Liu et al. (2015)

delay. The figure shows that the exponential time delay tends to be biased to the overestimation side in predicting FCP all the time.

The above comparison shows that the proposed models can provide a good estimation and prediction on the dataset with multiple releases. In this case, it is more suitable to assume the correcting time delay to be a Weibull distribution.

6.5 Summary

In this chapter, a framework of software reliability estimation based on fault detection and correction processes is developed, and parameter estimation problem in this situation is studied. Our approach takes into consideration the exact removal time interval of each fault. To describe this type of dataset, the fault removing matrix is defined. Our research is also extended to the software projects with multiple releases. The proposed modeling framework is applied to a real three-version test dataset. The results show that the proposed model framework with maximum likelihood estimation produces better parameter estimates and a reliable stochastic model. Also, the proposed model framework can provide a good estimation and prediction on the dataset with multiple releases.

Finally, it should point out that related to the proposed approach, there are quite a lot of research questions remaining to be further studied, such as dealing with the imperfectly debugging and faults' severities based on this approach,

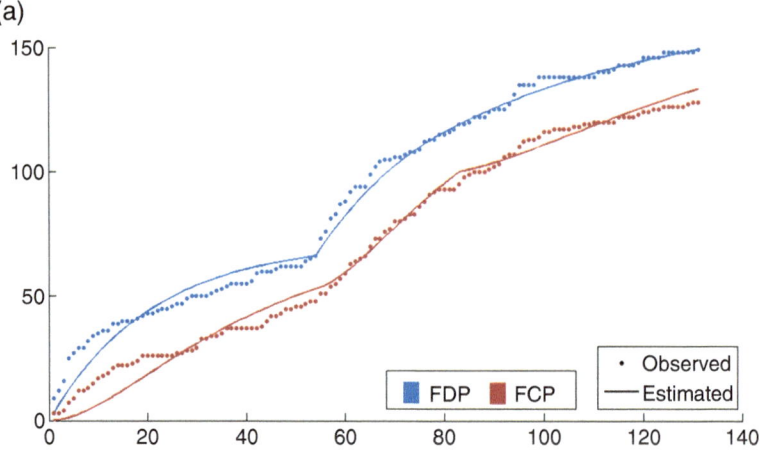

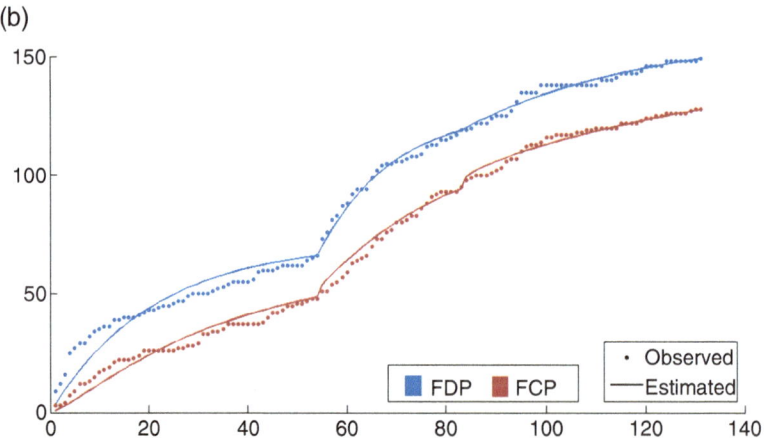

Fig. 6.5 Proposed model with multiple releases ©[2015] IEEE. (**a**) Exponential time delay. (**b**) Weibull time delay. (Reprinted, with permission, from Liu et al. 2015)

addressing the cost analysis, and incorporating the masked failure data into modeling. In future works, the proposed modeling framework with semigroup data will be extended with those issues and try to form more realistic assumptions.

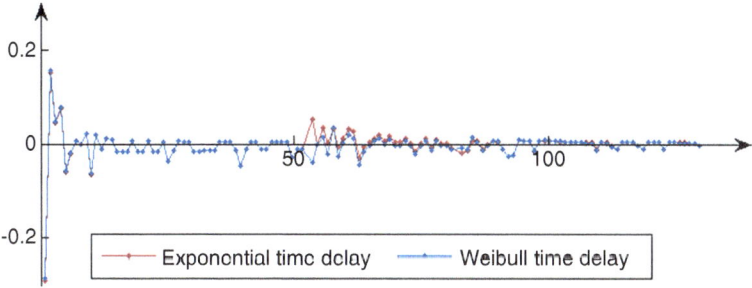

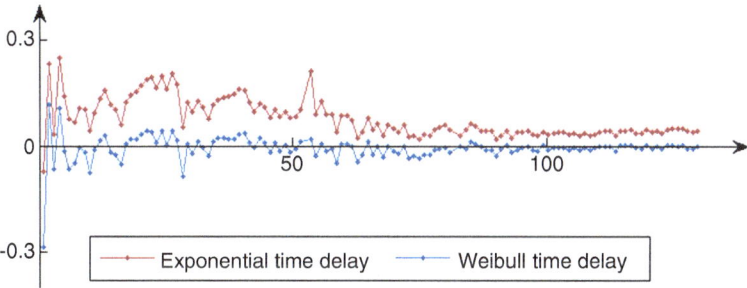

Fig. 6.6 Relative error of prediction with multiple released dataset ©[2015] IEEE. (Reprinted, with permission, from Liu et al. 2015)

References

Gaver, D. P., & Jacobs, P. A. (2014). Reliability growth by failure mode removal. *Reliability Engineering & System Safety, 130*, 27–32.

Hu, Q. P., Xie, M., Ng, S. H., & Levitin, G. (2007). Robust recurrent neural network modeling for software fault detection and correction prediction. *Reliability Engineering & System Safety, 92* (3), 332–340.

Huang, C. Y., & Huang, W. C. (2008). Software reliability analysis and measurement using finite and infinite server queueing models. *IEEE Transactions on Reliability, 57*(1), 192–203.

Huang, C. Y., & Huang, T. Y. (2010). Software reliability analysis and assessment using queueing models with multiple change-points. *Computers & Mathematics with Applications, 60*(7), 2015–2030.

Lawless, J. F. (2011). *Statistical models and methods for lifetime data* (Vol. 362). John Wiley & Sons.

Lin, C. T. (2011). Analyzing the effect of imperfect debugging on software fault detection and correction processes via a simulation framework. *Mathematical and Computer Modelling, 54*(11–12), 3046–3064.

Lin, C. T., & Li, Y. F. (2014). Rate-based queueing simulation model of open source software debugging activities. *IEEE Transactions on Software Engineering, 40*(11), 1075–1099.

Liu, Y., Xie, M., Yang, J., & Zhao, M. (2015). *A new framework and application of software reliability estimation based on fault detection and correction processes*. In IEEE international conference on Software Quality, 65–74.

Malaiya, Y. (1992). Predictability of software-reliability models. *IEEE Transactions on Reliability, 41*(4), 539–546.

Peng, R., Li, Y. F., Zhang, W. J., & Hu, Q. P. (2014). Testing effort dependent software reliability model for imperfect debugging process considering both detection and correction. *Reliability Engineering & System Safety, 126*, 37–43.

Ross, S. (1996). *Stochastic processes*. Wiley series in probability and mathematical statistics.

Schneidewind, N. F. (2001). *Modelling the fault correction process*. In Proceedings of 12th international symposium on software reliability engineering 185–190.

Wu, Y. P., Hu, Q. P., Xie, M., & Ng, S. H. (2008). Modeling and analysis of software fault detection and correction process by considering time dependency. *IEEE Transactions on Reliability, 56*(4), 629–642.

Xie, M., Hu, Q. P., Wu, Y. P., & Ng, S. H. (2007). A study of the modeling and analysis of software fault-detection and fault-correction processes. *Quality and Reliability Engineering International, 56*, 459–470.

Chapter 7
FDP and FCP with Four Types of Faults

This chapter incorporates different easiness of detection and correction of the software fault into the modeling of software fault detection process and fault correction process. In particular, the faults are classified as four types: (1) easy to detect and easy to correct; (2) easy to detect, but difficult to correct; (3) difficult to detect and easy to correct; and (4) difficult to detect and difficult to correct. The parameter estimation method is introduced, and the applications of the models are illustrated with real dataset. The optimal software release time policy is also studied.

7.1 Introduction

In order to guarantee the reliability of software, software usually needs to undergo a long testing phase before its release to market (Schneidewind 2009; Cinque et al. 2017). During software testing phase, software faults may be detected by the debuggers and finally corrected. During the last four decades, numerous software reliability growth models (SRGMs) have been proposed to track the reliability growth of software during testing, among which the nonhomogeneous Poisson process (NHPP) models are the most widely accepted.

Most NHPP models have assumed that a fault can be immediately corrected upon detection, that is, only fault detection process (FDP) is considered. Huang et al. (2003) presented a unified scheme of some NHPP models for describing the fault detection process. In order to take into account the uneven testing resource allocation during testing phase, some works have incorporated various types of the testing effort function into the NHPP models (Kuo et al. 2001; Huang 2005; Kapur et al. 2008). Lin and Huang (2008) incorporated the important concept of multiple change points into Weibull-type testing-effort functions. Zhang et al. (2016) first incorporated a multivariate function of FDR, which is bathtub-shaped, into the NHPP-based

R. Peng et al., *Software Fault Detection and Correction: Modeling and Applications*, SpringerBriefs in Computer Science, https://doi.org/10.1007/978-981-13-1162-8_7

SRGMs considering testing effort in order to further improve performance. Some researchers considered environmental change during software testing. Song and Chang (2007) proposed a new nonhomogeneous Poisson process (NHPP)-based software reliability model, with S-shaped growth curve for use during the software development process, and relate it to a fault detection rate function when considering random operating environments. Li and Pham (2017) proposed a software reliability model with consideration of the testing coverage. Some other researchers have incorporated the change point or phase transition into the software liability models (Lin 2012). In addition, some other models have tried to incorporate the imperfect debugging and the fault introduction process (Kapur et al. 2011; Li et al. 2015; Wang et al. 2015). Rawat and Goyal (2017) proposed a generic software reliability model for the agile process, taking permanent and transient faults into consideration. Ramasamy and Govindasamy (2008) incorporated the learning process into the software reliability modeling process. Dai et al. (2011) presented an innovative model and technology to realize the self-healing function under the real-time requirement.

During the last 10 years, some models have been proposed by considering the debugging delay, which is the time from the detection of a fault to the correction of the fault (Chatterjee and Shukla 2016). Actually, a detected fault needs to be reported, diagnosed, and finally corrected. Therefore, the debugging delay is not negligible. By incorporating the debugging delay, some researchers have proposed some paired models of fault detection process and fault correction process (FCP), where the fault correction process is modeled as a delayed process of the fault detection process. Some paired models of fault detection process and fault correction process are proposed in Lo and Huang (2006) and Xie et al. (2007) under different assumptions of debugging delay. Hu et al. (2007) modeled the FDP and FCP using a data-driven artificial neural network model. Yang and Li (2010) proposed a hybrid genetic algorithm (GA)-based algorithm which adopts the model mining technique to discover the correlation of failures and to obtain optimal model parameters. Huang and Huang (2008) incorporated the debugging delay and imperfect debugging into the software reliability modeling process based on a queueing model. Peng et al. (2014) proposed several testing effort-dependent models for both fault detection and correction processes. Peng and Zhai (2017) studied the software fault detection process and fault correction process considering two types of faults, where the dependent faults can be detected only after the leading faults are removed. Liu et al. (2016) studied a framework of software reliability models containing both information from software fault detection process and correction process.

Though these works have made the software reliability models one step closer to reality, all these works have assumed that all the faults have the same detection rate and the same distribution function for the debugging delay. In practice, some faults are easy to detect, whereas others are not. Debuggers have experiences that some faults may be due to grammar errors or simply typo; these faults are usually easy to detect. Some other faults may be due to space overflow or logical errors; these faults are usually hard to detect. Similarly, some fault may be very easy to correct, whereas others are difficult. In order to make the software reliability modeling further closer to reality, four types of faults are considered in this chapter: (1) easy to detect and correct, (2) easy to detect but difficult to correct, (3) difficult to detect but easy to correct, and (4) difficult to detect and correct. As a beginning, it is assumed that whether a fault is easy to detect and whether a fault is easy to correct are independent.

The remaining of this chapter is as follows. Section 7.2 proposes the model with incorporation of four types of fault model. Section 7.3 deals with the parameter estimation method. Section 7.4 illustrates the application of the proposed model with real dataset. Section 7.5 illustrates the application of the proposed model on the determination of the optimal release time. Section 7.6 concludes.

7.2 The Modeling Process

The total number of faults in the software is assumed to be Poisson distributed, and the expected value is denoted as a. The proportion of faults that are easy to detect is denoted as p. The proportion of faults that are easy to correct is denoted as q. As the fault that is easy to detect may not be easy to correct, it is assumed that the easiness to detect and the easiness to correct are independent. Therefore, the expected number of faults that are easy to detect and correct is apq, the expected number of faults that are easy to detect but difficult to correct is $ap(1-q)$, the expected number of faults that are difficult to detect but easy to correct is $a(1-p)q$, and the expected number of faults that are difficult to detect and correct is $a(1-p)(1-q)$. The fault detection rate for the faults that are easy to detect is b_1, and the fault detection rate for the faults that are difficult to detect is $b_2 < b_1$. The fault correction time for the faults are assumed to be exponential distributed, and the parameter of the exponential distribution is termed as the fault correction rate. The correction rate for the faults that are easy to correct is c_1, and the correction rate for the faults that are difficult to correct is $c_2 < c_1$. The modeling of software reliability growth is usually characterized by the mean value function, which is the expected number of detected or correct fault up to time t of the testing process. In our cases, there are four types of faults: (1) easy to detect and correct, (2) easy to detect but difficult to correct, (3) difficult to detect but easy to correct, and (4) difficult to detect and correct. Therefore, the mean value functions for their detection process and correction process can be denoted as $m_{di}(t)$ and $m_{ci}(t)$, where "d" refers to "detection," "c" refers to "correction," and $i = 1, 2, 3, 4$ denotes the type of faults of concern.

7.2.1 Fault Detection Process

We assume that FDP for each type of fault follows a NHPP, and the expected number of fault detected during $(t, t + \Delta t]$ is proportional to the number of undetected faults at time t. Thus, we have

$$\frac{dm_{d1}(t)}{dt} = b_1(apq - m_{d1}(t)), \tag{7.1}$$

$$\frac{dm_{d2}(t)}{dt} = b_1(ap(1-q) - m_{d2}(t)), \tag{7.2}$$

$$\frac{dm_{d3}(t)}{dt} = b_2(a(1-p)q - m_{d3}(t)), \tag{7.3}$$

$$\frac{dm_{d4}(t)}{dt} = b_2(a(1-p)(1-q) - m_{d4}(t)). \tag{7.4}$$

It is easy to obtain that

$$m_{d1}(t) = apq(1 - e^{-b_1 t}), \tag{7.5}$$

$$m_{d2}(t) = ap(1-q)(1 - e^{-b_1 t}), \tag{7.6}$$

$$m_{d3}(t) = a(1-p)q(1 - e^{-b_2 t}), \tag{7.7}$$

$$m_{d4}(t) = a(1-p)(1-q)(1 - e^{-b_2 t}). \tag{7.8}$$

Combining the four types of faults together, the mean value function for fault detection process can be obtained as

$$m_d(t) = ap(1 - e^{-b_1 t}) + a(1-p)(1 - e^{-b_2 t}) \tag{7.9}$$

7.2.2 Fault Correction Process

FCP can be regarded as a delayed process of FDP for the four types of faults. As fault type 1 and fault type 3 are both easy to correct, we model them together. Similarly, we model fault type 2 and fault type 4 together. It is easy to have

$$\frac{d(m_{c1}(t) + m_{c3}(t))}{dt} = c_1(m_{d1}(t) + m_{d3}(t) - m_{c1}(t) - m_{c3}(t)), \tag{7.10}$$

$$\frac{d(m_{c2}(t) + m_{c4}(t))}{dt} = c_2(m_{d2}(t) + m_{d4}(t) - m_{c2}(t) - m_{c4}(t)), \tag{7.11}$$

Substituting (7.5)–(7.8) into (7.10) and (7.11), we have

$$m_{c1}(t) + m_{c3}(t) = \frac{apqc_1}{c_1 - b_1}\left(e^{-c_1t} - e^{-b_1t}\right) + \frac{a(1-p)qc_1}{c_1 - b_2}\left(e^{-c_1t} - e^{-b_2t}\right)$$

$$+ aq(1 - e^{-c_1t}), \tag{7.12}$$

$$m_{c2}(t) + m_{c4}(t) = \frac{ap(1-q)c_2}{c_2 - b_1}\left(e^{-c_2t} - e^{-b_1t}\right)$$

$$\mid \frac{a(1-p)(1-q)c_2}{c_2 - b_1}\left(e^{-c_2t} - e^{-b_2t}\right)$$

$$+ a(1-q)(1 - e^{-c_2t}), \tag{7.13}$$

Combining (7.12) and (7.13) gives

$$m_c(t) = \frac{apqc_1}{c_1 - b_1}\left(e^{-c_1t} - e^{-b_1t}\right) + \frac{a(1-p)qc_1}{c_1 - b_2}\left(e^{-c_1t} - e^{-b_2t}\right)$$

$$+ aq(1 - e^{-c_1t}) + \frac{ap(1-q)c_2}{c_2 - b_1}\left(e^{-c_2t} - e^{-b_1t}\right)$$

$$+ \frac{a(1-p)(1-q)c_2}{c_2 - b_1}\left(e^{-c_2t} - e^{-b_2t}\right) + a(1-q)(1 - e^{-c_2t}), \tag{7.14}$$

For convenience of discussion, the model described by (7.9) and (7.14) is termed as M1.

7.2.3   Some Special Cases

In order to show the effects of considering four types of faults, we need to compare with the models considering two types of faults with the same detection rate but different correction rates (M2), two types of faults with the same correction rate but different detection rates (M3), and only one type of faults (M4).

Let $b_1 = b_2 = b$ and $p = 1$, and still keep $c_1 > c_2$; M2 can be obtained as

$$m_d(t) = a(1 - e^{-bt}),$$

$$m_c(t) = \frac{aqc_1}{c_1 - b}\left(e^{-c_1t} - e^{-bt}\right) + aq(1 - e^{-c_1t}) + \frac{a(1-q)c_2}{c_2 - b}\left(e^{-c_2t} - e^{-b_1t}\right)$$

$$+ \frac{a(1-q)c_2}{c_2 - b}\left(e^{-c_2t} - e^{-bt}\right) + a(1-q)(1 - e^{-c_2t}).$$

Keep $b_1 > b_2$, and let $c_1 = c_2 = c$ and $q = 1$; M3 can be obtained as

$$m_d(t) = ap(1 - e^{-b_1 t}) + a(1 - p)(1 - e^{-b_2 t}),$$

$$m_c(t) = \frac{apc_1}{c - b_1}(e^{-ct} - e^{-b_1 t}) + \frac{a(1 - p)c}{c - b_2}(e^{-ct} - e^{-b_2 t}) + a(1 - e^{-ct}).$$

Let $b_1 = b_2 = b$, $c_1 = c_2 = c$, and $q = 1$; M4 can be expressed as

$$m_d(t) = a(1 - e^{-bt}),$$

$$m_c(t) = \frac{ac}{c - b}(e^{-ct} - e^{-bt}) + a(1 - e^{-ct}).$$

Note that M4 is the same as the model obtained in Lo and Huang (2006), Xie et al. (2007), and Wu et al. (2008) under the assumptions of constant detection rate and exponential debugging delay.

7.3   Parameter Estimation

In practice, the model parameters need to be estimated by fitting the model to the collected data of fault detection numbers and fault correction numbers during different time intervals. A typical approach is the least squares method, which minimizes the mean squared error (MSE) between the theoretical model and the actual observation. In our case, the MSE can be calculated as

$$MSE = \frac{1}{2}(MSE_d + MSE_c) = \frac{1}{2n}\sum_{i=1}^{n}\left[(m_d(t_i) - m_{d,i})^2 + (m_c(t_i) - m_{c,i})^2\right],$$

$$(7.15)$$

In addition, the root for MSE can be obtained as

$$RMSE = \sqrt{MSE}.$$

where $m_{d,\ i}$ and $m_{c,\ i}$ are the observed cumulative numbers of detected faults and corrected faults at time t_i, $i = 1, \ldots, n$.

7.4   Illustrative Examples

The dataset is from the System T1 data of the Rome Air Development Center (RADC) (Musa et al. 1987). This dataset is widely used, and it contains both fault detection data and fault correction data. The cumulative numbers of detected faults and corrected faults during the first 21 weeks are shown in Table 7.1.

Table 7.1 The dataset: System T1

Weeks	Computer time (CPU hours)	Cumulative number of detected faults (m_d)	Cumulative number of corrected faults (m_r)
1	4	2	1
2	8.3	2	2
3	10.3	2	2
4	10.9	3	3
5	13.2	4	4
6	14.8	6	4
7	16.6	7	5
8	31.3	16	7
9	56.4	29	13
10	60.9	31	17
11	70.4	42	18
12	78.9	44	32
13	108.4	55	37
14	130.4	69	56
15	169.9	87	75
16	195.9	99	85
17	220.9	111	97
18	252.3	126	117
19	282.3	132	129
20	295.1	135	131
21	300.1	136	136

During the time span, 300.1 h of computer time were consumed, and 136 faults were detected and corrected.

The proposed models are fitted to the dataset by the least squared method. The estimated model parameters by the four models for dataset are given in Table 7.2.

As can be noticed from Table 7.2, the estimated parameter a (the total number of faults) in the M1 is the closest to 188, which is the number of detected faults after 3-year testing, as reported in Kapur and Younes (1995). The RMSE is also considerably lower for $M1$. It shows that considering different detection rates and correction rates does have some advantage.

Figure 7.1 gives a contrast diagram of all cases for *RMSE*. It is obvious that the *RMSE* value of the model $M1$ is minimal and the $M1$ fitting is optimal.

Table 7.2 The estimated model parameters for dataset

Model	M1	M2	M3	M4
Remark	$b_1 > b_2, c_1 > c_2$	$b_1 = b_2(c_1 > c_2)$	$c_1 = c_2(b_1 > b_2)$	$b_1 = b_2, c_1 = c_2$
a	192.8715	159.7669	157.0116	155.6930
p	0.0179	1	0.0049	1
b_1	0.0047	0.0053	0.0057	0.0056
b_2	0.0039	0.0053	0.0054	0.0056
q	0.0385	0.0586	1	1
c_1	0.0398	0.0283	0.0251	0.0253
c_2	0.0258	0.0250	0.0251	0.0253
RMSE	5.9755	7.9891	8.1860	8.3626

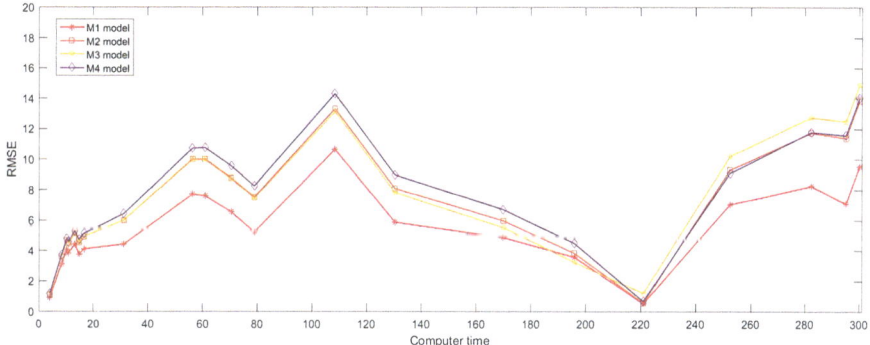

Fig. 7.1 The RMSE charts of M1–M4 for dataset

7.5   Software Release Policies

The determination of the best release time is a key decision of the software project, which generally considers strategies based on cost and reliability (Ho et al. 2008). In this section, we also study the optimal release strategy considering the cost and reliability based on our proposed model.

7.5.1   Software Release Policy Based on Reliability Criterion

Software is usually released when a reliability target is achieved. It is reasonable to stop testing when a pre-specified proportion of faults are removed. We use T to denote the length of testing and consider the ratio of cumulative removed faults to the initial faults in the software system as the reliability criterion (Huang and Lin 2006)

$$R_1(T) = \frac{m_c(T)}{a}$$

$$T_1 = m_c^{-1}(a \cdot R_1)$$

Another criterion is the software reliability, which is defined as the probability that no faults occur during time interval $[T, T + \Delta T]$ given that the software is released at time T. Considering that the reliability status of software generally does not change in operational phase, the reliability function is

$$R_2(\Delta T \mid T) = \exp[-\lambda_d(T)\Delta T]$$

where $\lambda_d(T)$ is the instantaneous failure intensity at time T. For a given target R_2 for $R_2(\Delta T|T)$, the time for the software to reach R_2 can be solved as $T_2 = \min_T \{T : R_2(\Delta T |T) \geq R_2\}$.

7.5.2 Software Release Policy Based on Cost Criterion

For a mean value function $m_d(t)$ and $m_c(t)$, the following cost model is frequently used (Peng and Zhai 2017):

$$C(T) = c_1 m_c(T) + c_2[m_d(\infty) - m_c(T)] + c_3 T$$

where c_1 is the expected cost of removing a fault during testing, c_2 is the expected cost of removing a fault in the field, and c_3 is the expected cost per unit time of testing. The cost of removing a fault in field is generally greater than that during testing; thus, we assume $c_2 > c_1$.

In case where the easy- or difficult-detected faults and easy- or difficult-corrected faults are incorporated, the following cost model can be constructed:

$$\begin{aligned}
C(T) = &\, k_1[m_{d1}(T) + m_{d2}(T)] + k_2[m_{d3}(T) + m_{d4}(T)] + k_3[m_{c1}(T) + m_{c3}(T)] \\
&+ k_4[m_{c2}(T) + m_{c4}(T)] + k_5[m_{d1}(\infty) + m_{d3}(\infty) - m_{c1}(T) - m_{c3}(T)] \\
&+ k_6[m_{d2}(\infty) + m_{d4}(\infty) - m_{c2}(T) - m_{c4}(T)] + k_7 T
\end{aligned}$$

where $m_{d1}(T) + m_{d2}(T)$ are the total number of detected faults which are easily to detect at the time of release T, $m_{d3}(T) + m_{d4}(T)$ are the total number of detected faults which are difficult to detect at the time of release T, $m_{c1}(T) + m_{c3}(T)$ are the total number of corrected faults which are easy to correct at the time of release T, and $m_{c2}(T) + m_{c4}(T)$ are the total number of corrected faults which are difficult to correct at the time of release T. Note that, to be more general, the model allows the cost of detecting each easy detectable fault k_1 to be different from the cost of detecting each difficult detectable fault k_2 and the cost of correcting each easy correctable fault k_3 to be different from each difficult correctable fault k_4. $m_{d1}(\infty) + m_{d3}(\infty) - m_{c1}(T) - m_{c3}(T)$ is the number of uncorrected faults which are easy to correct and which include both the undetected faults $m_{d1}(\infty) + m_{d3}(\infty) - m_{d1}(T) - m_{d3}(T)$ and the detected but not corrected faults $m_{d1}(T) + m_{d3}(T) - m_{c1}(T) - m_{c3}(T)$. $m_{d2}(\infty) + m_{d4}(\infty) - m_{c2}(T) - m_{c4}(T)$ is the number of uncorrected faults that are difficult to correct, which includes both the undetected faults $m_{d2}(\infty) + m_{d4}(\infty) - m_{d2}(T) - m_{d4}(T)$ and the detected but not corrected faults $m_{d2}(T) + m_{d4}(T) - m_{c2}(T) - m_{c4}(T)$. k_5 is the expected cost of removing a fault that is easy to correct in the field. k_6 is the expected cost of removing a fault that is difficult to correct in the field. k_7 is the expected cost per unit time of testing. According to the definitions, it is reasonable to assume that

$$k_2 > k_1 > 0, k_4 > k_3 > 0, k_6 > k_5 > 0.$$

By minimizing the cost model with respect to T, the optimal release time T_c under the proposed framework can be obtained.

Theorem 1 The minimum value of the cost function $C(T)$ exists.

Proof The cost function is a continuous elementary function, so it is continuous in $[0, T]$. By the minimax theorem in mathematics, the minimum value of $C(T)$ in $[0, T]$ exists. As $\mathrm{Lim}_{T \to \infty} C'(T) = k_7$, by the limit definition and the limit property, there is a sufficiently large $\hat{T}$ that makes $C'(T) > 0$ for any $T > \hat{T}$. Therefore, the minimal value of $C(T)$ in $[0, \infty)$ is the minimum value $C(T)$ in $[0, T]$.

7.5.3   Software Release Policy Based on Mixed Criterion

When both software reliability and the total cost are considered, we determine the optimal release time T^* that minimizes the total cost under the reliability constraint. Accordingly, the problem can be formulated as
   Minimize:

$$C(T) = k_1[m_{d1}(T) + m_{d2}(T)] + k_2[m_{d3}(T) + m_{d4}(T)] + k_3[m_{c1}(T) + m_{c3}(T)]$$
$$+ k_4[m_{c2}(T) + m_{c4}(T)] + k_5[m_{d1}(\infty) + m_{d3}(\infty) - m_{c1}(T) - m_{c3}(T)]$$
$$+ k_6[m_{d2}(\infty) + m_{d4}(\infty) - m_{c2}(T) - m_{c4}(T)] + k_7 T$$

subject to $R_1(T) = \frac{m_c(T)}{a}$ or $R_2(\Delta T | T) = \exp[-\lambda_d(T)\Delta T]$.
   Similar to theorem 1, if $C'(T) > 0$ for any $T > TT > \max(T_1, T_2)$, the optimal release time is the lowest point of $C(T)$ in $[T_1, TT]$ or $[T_2, TT]$.

7.5.4   Numerical Examples for Software Release Policy

For illustration, we consider the model $M1$ in Sect. 7.3. The parameters for model $M1$ are $a = 192.8715$, $p = 0.0179$, $b_1 = 0.0047$, $b_2 = 0.0039$, $q = 0.0385$, $c_1 = 0.0398$, and $c_2 = 0.0258$.

Assume other parameters are $k_1 = 50$, $k_2 = 100$, $k_3 = 200$, $k_4 = 300$, $k_5 = 1000$, $k_6 = 1500$, $k_7 = 30$, $\Delta T = 12$, $R_1(T) = 0.95$, and $R_2(T) = 0.95$; substituting these parameters into the cost function and the reliability functions, the optimal release time is obtained.

Let $R_1(T) = 0.95$, T_1 can be obtained as 807.4737. Let $R_2(T) = 0.95$, T_2 can be obtained as 1323. The minimum value of the cost function $C(T_3) = 110,700$ and $T_3 = 892.4593$. Furthermore, the curves for $R_1(T)$, $R_2(T)$, and $C(T)$ are drawn in Fig. 7.2. As can be seen from Fig. 7.2, when considering $R_1(T)$ and $C(T)$, the optimal release time equals to the minimum value point of cost function, that is, T_3. Thus, the optimal release time is $T^* = 892.4593$, and the corresponding cost is $C(T^*) = 110,700$. When considering $R_2(T)$, $C(T)$ is the optimal release time equals to T_2. Thus, the optimal release time $T^* = 1323$, and the corresponding cost is $C(T^*) = 117,360$.

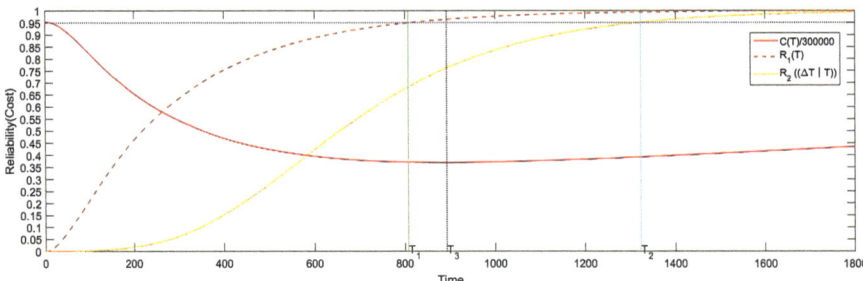

Fig. 7.2 The diagram of the cost function and the reliability function for the model M1

7.6   Summary

In this chapter, a software reliability model is proposed considering four types of faults, that is, (1) easy to detect and correct, (2) easy to detect but difficult to correct, (3) difficult to detect but easy to correct, and (4) difficult to detect and correct. In order to show the effects of considering these four types of faults, we also derive the models considering two types of faults with the same detection rate but different correction rates, two types of faults with the same correction rate but different detection rates, and only one type of faults. Real dataset is adopted to illustrate the applications, and the results show the advantage of our model. In order to illustrate the practical use of our model, the optimal release policy is investigated.

In this chapter, whether a fault is easy to correct or not is assumed to be irrelevant to whether it is easy to detect. In the future, the dependency between the easiness to detect and the easiness to correct can be investigated. Another future work is to incorporate the fault introduction effects as well. Say, it can be assumed that more faults tend to be introduced when correcting the faults which are difficult to correct.

References

Chatterjee, S., & Shukla, A. (2016). Modeling and analysis of software fault detection and correction process through weibull-type fault reduction factor, change point and imperfect debugging. *Arabian Journal for Science and Engineering, 41*(12), 5009–5025.

Cinque, M., Cotroneo, D., Pecchia, A., Pietrantuono, R., & Russo, S. (2017). Debugging-workflow-aware software reliability growth analysis. *Software Testing Verification & Reliability, 27*(7), e1638.

Dai, Y. S., Xiang, Y. P., & Li, Y. F. (2011). Consequence oriented self-healing and autonomous diagnosis for highly reliable systems and software. *IEEE Transactions on Reliability, 60*(2), 369–380.

Ho, J., Fang, C., & Huang, Y. (2008). The determination of optimal software release times at different confidence levels with consideration of learning effects. *Software Testing Verification & Reliability, 18*(4), 221–249.

Hu, Q. P., Xie, M., Ng, S. H., & Levitin, G. (2007). Robust recurrent neural network modeling for software fault detection and correction prediction. *Reliability Engineering & System Safety, 92*(3), 332–340.

Huang, C. Y. (2005). Performance analysis of software reliability growth models with testing-effort and change-point. *Journal of Systems and Software, 76*(2), 181–194.

Huang, C. Y., & Huang, W. C. (2008). Software reliability analysis and measurement using finite and infinite server queueing models. *IEEE Transactions on Reliability, 57*(1), 192–203.

Huang, C. Y., & Lin, C. T. (2006). Software reliability analysis by considering fault dependency and debugging time lag. *IEEE Transactions on Reliability, 55*(3), 436–450.

Huang, C. Y., Lyu, M. R., & Kuo, S. Y. (2003). A unified scheme of some nonhomogenous Poisson process models for software reliability estimation. *IEEE Transactions on Software Engineering, 29*(3), 261–269.

Kapur, P. K., & Younes, S. (1995). Software reliability growth model with error dependency. *Microelectronics Reliability, 35*(2), 273–278.

Kapur, P. K., Goswami, D. N., Bardhan, A., & Singh, O. (2008). Flexible software reliability growth model with testing effort dependent learning process. *Applied Mathematical Modelling, 32*(7), 1298–1307.

Kapur, P. K., Pham, H., Anand, S., & Yadak, K. (2011). A unified approach for developing software reliability growth models in the presence of imperfect debugging and error generation. *IEEE Transactions on Reliability, 60*(1), 331–340.

Kuo, S. Y., Huang, C. Y., & Lyu, M. R. (2001). Framework for modeling software reliability, using various testing-efforts and fault-detection rates. *IEEE Transactions on Reliability, 50*(3), 310–320.

Li, Q., & Pham, H. (2017). NHPP software reliability model considering the uncertainty of operating environments with imperfect debugging and testing coverage. *Applied Mathematical Modeling, 51*, 68–85.

Li, Q., Li, H., & Lu, M. (2015). Incorporating S-shaped testing-effort functions into NHPP software reliability model with imperfect debugging. *Journal of Systems Engineering and Electronics, 26*(1), 190–207.

Lin, C. T. (2012). Enhancing the accuracy of software reliability prediction through quantifying the effect of test phase transitions. *Applied Mathematics and Computation, 219*(5), 2478–2492.

Lin, C. T., & Huang, C. Y. (2008). Enhancing and measuring the predictive capabilities of testing-effort dependent software reliability models. *Journal of Systems and Software, 81*(6), 1025–1038.

Liu, Y., Li, D., Wang, L., & Hu, Q. (2016). A general modeling and analysis framework for software fault detection and correction process. *Software Testing Verification & Reliability, 26*(5), 351–365.

Lo, J. H., & Huang, C. Y. (2006). An integration of fault detection and correction processes in software reliability analysis. *Journal of Systems and Software, 79*, 1312–1323.

Musa, J. D., Iannino, A., & Okumoto, K. (1987). Software reliability, measurement. In *Prediction and application*. New York: McGraw-Hill.

Peng, R., & Zhai, Q. (2017). Modeling of software fault detection and correction processes with fault dependency. *Eksploatacja i Niezawodnosc-Maintenance and Reliability, 19*(3), 467–475.

Peng, R., Li, Y. F., Zhang, W. J., & Hu, Q. P. (2014). Testing effort dependent software reliability model for imperfect debugging process considering both detection and correction. *Reliability Engineering and System Safety, 126*, 37–43.

Ramasamy, S., & Govindasamy, G. (2008). A software reliability growth model addressing learning. *Journal of Applied Statistics, 35*(10), 1151–1168.

Rawat, S., & Goyal, N. (2017). Software reliability growth modeling for agile software development. *International Journal of Applied Mathematics and Computer Science, 27*(4), 777–783.

Schneidewind, N. (2009). Integrating testing with reliability. *Software Testing Verification & Reliability, 19*(3), 175–198.

Song, K. Y., & Chang, I. H. (2007). An NHPP software reliability model with S-shaped growth curve subject to random operating environments and optimal release time. *Applied Science-Basel, 7*(12), 1304.

Wang, J. Y., Wu, Z. B., Shu, Y. J., & Zhang, Z. (2015). An imperfect software debugging model considering log-logistic distribution fault content function. *Journal of Systems and Software, 100*, 167–181.

Wu, Y. P., Hu, Q. P., Xie, M., & Ng, S. H. (2008). Modeling and analysis of software fault detection and correction process by considering time dependency. *IEEE Transactions on Reliability, 56*(4), 629–642.

Xie, M., Hu, Q. P., Wu, Y. P., & Ng, S. H. (2007). A study of the modeling and analysis of software fault-detection and fault-correction processes. *Quality and Reliability Engineering International, 23*, 459–470.

Yang, B., & Li, X. (2010). A generic data-driven software reliability model with model mining technique. *Reliability Engineering & System Safety, 95*(6), 671–678.

Zhang, J., Lu, Y., Yang, S., & Xu, C. (2016). NHPP-based software reliability model considering testing effort and multivariate fault detection rate. *Journal of Systems Engineering and Electronics, 27*(1), 260–270.